71st Conference
on Glass Problems

71st Conference on Glass Problems

*A Collection of Papers Presented at the
71st Conference on Glass Problems
The Ohio State University, Columbus, Ohio
October 19–20, 2010*

Edited by
Charles H. Drummond, III

A John Wiley & Sons, Inc., Publication

Published by John Wiley & Sons, Inc., Hoboken, New Jersey.
Published simultaneously in Canada.

For general information on our other products and services or for technical support, please contact our Customer Care Department within the United States at (800) 762-2974, outside the United States at (317) 572-3993 or fax (317) 572-4002.

Wiley also publishes its books in a variety of electronic formats. Some content that appears in print may not be available in electronic format. For information about Wiley products, visit our web site at www.wiley.com.

Library of Congress Cataloging-in-Publication Data is available.

ISBN: 978-1-118-05996-8
ISBN: 978-1-118-09560-7 (special edition)
ISSN: 0196-6219

oBook ISBN: 978-1-118-09534-8
ePDF ISBN: 978-1-118-10642-6

Printed in the United States of America.

10 9 8 7 6 5 4 3 2 1

Contents

REFRACTORIES AND RECYCLING

CONTROLS AND RAW MATERIALS

Foreword

The conference was sponsored by the Department of Materials Science and Engineering at The Ohio State University.

The director of the conference was Charles H. Drummond, III, Associate Professor, Department of Materials Science and Engineering, The Ohio State University.

Daniel Kramer, Industry Liaison Executive, College of Engineering, The Ohio State University, gave the welcoming address.

The themes and chairs of the four half-day sessions were as follows:

Glass Melting
Ruud Beerkens, TNO Glass Technology—Glass Group, Eindhoven, The Netherlands, Tom Dankert, O-I, Toledo OH and Larry McCloskey, Toledo Engineering, Toledo OH

Glass Glass Science, Defects and Safety
Philip Tucker, Johns Manville, Denver CO, Elmer Sperry, Libbey Glass, Toledo OH and Martin H. Goller, Corning, Corning NY

Glass Refractories and Recycling
Matthew Wheeler, RHI Monofrax, Batavia OH, Jack Miles and H. C. Starck, Coldwater MI

Glass Controls and Raw Materials
Warren F. Curtis, PPG Industries, Pittsburgh PA and Glenn Neff, Glass Service, Stuart FL

Preface

In the tradition of previous conferences, started in 1934 at the University of Illinois, the papers presented at the 71st Annual Conference on Glass Problems have been collected and published as the 2011 edition of The Collected Papers.

The manuscripts are reproduced as furnished by the authors, but were reviewed prior to presentation by the respective session chairs. Their assistance is greatly appreciated. C. H. Drummond did minor editing with further style editing by The American Ceramic Society. The Ohio State University is not responsible for the statements and opinions expressed in this publication.

CHARLES H. DRUMMOND, III

Columbus, OH
January 2011

Acknowledgments

It is a pleasure to acknowledge the assistance and advice provided by the members of Program Advisory Committee in reviewing the presentations and the planning of the program:

Ruud G. C. Beerkens—TNO
Warren Curtis—PPG Industries
Tom Dankert—O-I
Martin H. Goller—Corning
Jack Miles—H. C. Stark
Glenn Neff—Glass Service
Elmer Sperry—Libbey Glass
Philip Tucker—Johns Manville
Carsten Weinhold—Schott
Matthew Wheeler—RHI Monofrax
Dan Wishnick—Siemens

Glass Melting

RECENT DEVELOPMENTS OF BATCH AND CULLET PREHEATING IN EUROPE—PRACTICAL EXPERIENCES AND IMPLICATIONS

Philipp Zippe
Zippe Industrieanlagen GmbH
Wertheim, Germany

ABSTRACT

Batch & Cullet preheating itself is not new. First installations in the glass industry go back until the early 80ies of the last century. But, even after first successes and significant energy savings (~14-18%), the demand was not increasing, due to low energy prices, investment costs, risk aversion of decision makers, and also certain shortcomings of existing systems. However, in the last years the request for this technology increased significantly and the first new installations were made. One major challenge in first generation systems was the evaporation of batch moisture in the preheater, which resulted in batch clogging and maintenance efforts and restricted the application to batch with cullet ratios above approximately 50 per cent.

In order to improve the existing technology, Zippe undertook considerable R & D activities and finally initiated a pilot project, together with a major European container glass producer and a leading company in furnace design to test and prove the superiority of the modern, 2nd generation system.[1] In the meantime, in 2010, also another batch & cullet preheater (350 mt/d) was installed and first results are available. The paper will deal with these new experiences made and shall elaborate under which circumstances modern batch preheating shall be taken into account to save energy and therefore energy costs.

ENVIRONMENTAL CONTEXT

The productivity of the glass industry underwent significant improvements in the last decades. Compared to 1970, productivity in Germany approximately tripled and grew from 17.100 € to 54.800 €.[2] While it seems that a lot has been done to improve efficiency, there are still potentials and also necessities for further improvements. In 2005 the EU implemented the so called ETS (Emission Trading Scheme) that comprises all 27 member states and forces energy intensive industries (in total 12000 production plants)- as the glass industry- to focus on CO_2 emissions. Each company that is listed in the national allocation plan (Nationaler Allokationsplan) needs to disclose their yearly emissions which are then compared to a given and approved target. From the next trading period on in 2013, the values will be compared to reference glass factories with BAT (Best Available Technique). While also considering eventual already realized improvements, the factory is given a target emission, for example 510 kg CO_2/mt of flat glass that needs to be fulfilled.[3] This target is reduced in certain periods such that the factory is obliged to undertake further efficiency improvements. If the factory emits more

---

[1] The combined research project "PRECIOUS" started in 2006 and was supported by the German Environment Ministry DBU (Deutsche Bundesstiftung Umwelt). The theoretical research was done in cooperation with a major German technical university. The tests are accompanied by a reputable research organisation.
The start of the "PRECIOUS" project was already presented by Dr. Ann-Kathrin Glüsing: "Preheating Devices for Future Glass Making, a 2nd generation." 67th Conference on Glass Problems, Columbus Ohio, 30th Oct.- 1st Nov.
[2] Neckermann Gerhard, Wessels Hans. Die Glasindustrie-ein Branchenbild. Deutsches Institut für Wirtschaftsforschung. Heft 95, 1987. EUROSTAT: The european department of statistics (Brussels).
http://epp.eurostat.ec.europa.eu/portal/page/portal/eurostat/home/. The indicator for productivity is indic_sb v91110. Defined as gross value added per employee per year (€).
[3] Nationaler Allokationsplan Deutschland from 28.06.2006. obtained from (28.07.2010):
http://www.bmu.de/files/emissionshandel/downloads/application/pdf/nap_2008_2012.pdf. Page 53.

CO_2 than foreseen for free allocation of CO_2 permits, it needs to purchase so called CO_2-certificates which are traded at the European Energy Exchange in Leipzig, Germany. The prices in May 2010 fluctuated between 15 and 16 €/ton CO_2. Prices are expected to increase significantly in the next years, especially after the beginning of the next trading period, starting in 2013, when the quantity of certificates will be shortened, the benchmark values will be tightened and more of the certificates will also be auctioned. The aim of this market-based instrument is to enforce efficiency improvements by implementing incentives to firms to invest in pollution-minimizing (CO_2-lean) technologies. The incentives for firms are twofold. In order to avoid the necessity to purchase CO_2- certificates, firms may invest in emission-reducing technology, such as waste heat recovery systems. Furthermore, if a firm reduces its greenhouse gas emissions, it is allowed to sell non-needed allowances on the market which means an additional incentive to invest in "green technology."

Energy prices remain a concern for the European glass industry. Between 2005 and 2009, the gas prices for industrial users in the EU-27 raised from 6.0 to 9.4 €/GJ, which constitutes a price increase of almost 57%. In 1998 the respective price in the EU-15 was at level of 4.0 €/GJ.[4] In Germany- the biggest glass market in the EU- prices in 2007 remained at a level of more than 12 €/GJ and only slightly decreased in 2009 to about 11 €/GJ.

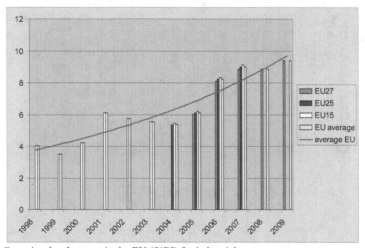

Graph 1 : Gas price development in the EU (€/GJ) for industrials users

These high prices and also the expectation of further increasing prices in the future, emphasized the necessity of the glass industry to invest in energy efficient technologies. However, for the melting of glass with today's technology, there are physical limits, which are almost reached.

---

[4] EUROSTAT- The statistical institute of the European Union, Brussels. Data obtained on 11.06.2010 from http://epp.eurostat.ec.europa.eu/portal/page/portal/eurostat/home/.

Table 1: development of key figures of the glass melting process[5]

year	1928	1968	1990	1998
specific heat demand (kWh/mt)	5600	2600	1550	1100
throughput (t/m²*d)	0,2	1,1	3,0	3,5
furnace lifetime(d)	300	2100	3000	4500
melting temperature(°C)	1370	1450	1500	1500
recycling ratio (%)	10	20	60	80
CO_2 emission (kg/mt)	1340	700	400	270

The specific energy demand for 1 ton glass is today about one fifth than it was in 1928. Also the throughput and the furnace lifetime increased dramatically. The emissions are today about one fifth compared to the beginning of industrial glass making with regenerative melting furnaces around 1928. Through increased utilization of cullet as a batch ingredient, energy can be saved. As commonly known, through the additional input of 10%, approx. 2-3% of melting energy can be saved. However, in Europe, the additional utilization of cullet is limited in many countries, since recycling ratios are already on a high level and the availability of reusable cullet constricted. Respective ratios in Switzerland are at about 95%, in Germany at about 90%.

Taking into consideration calculations from CONRADT, the theoretical process heat demand- which constitutes the minimal possible heat demand- is reached at a value of 920 kWh/t.[6] Looking at an industry average value of about 1000 kWh/t in 2003, it can be seen that potentials in glass melting for saving energy are limited, as the efficient frontier with today's technology is almost reached.

---

[5] Conradt Reinhard (2003)- HVG-Mitteilungen 2037, 1-6 (2003).
[6] Conradt Reinhard (2003)- HVG-Mitteilungen 2037, 1-6 (2003).

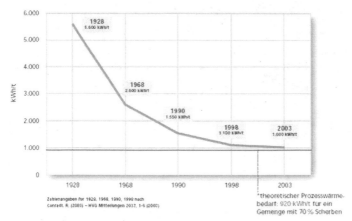

Graph 2: development of heat demand for glass melting process

Thus, to improve the energy efficiency in the glass industry to a reasonable and considerable extent, there need to be other starting points.

In the following, an overview of the energy flow of an existing container glass furnace in Europe can be seen. This furnace was taken from a best-practice-study and shows one of the most energy-efficient furnaces. At a cullet level of 84 per cent, the specific energy consumption amounts up to 3.62 GJ/ton. About 2.3 per cent of the total energy is needed as reaction energy, and about 45 per cent as heat content for the glass melt. Also the wall losses of about 17 per cent are considerable. However, the impact of additional insulation is limited. Thicker and stronger insulation would make the construction of the furnace more complicated and also have disadvantages concerning safety and operation of the furnace, as it would make the surveillance and the control of critical parts more difficult.[7] As wall losses of the regenerator and the energy needed for water evaporation are of minor quantity and can hardly be changed, it seems obvious to focus on the energy losses via flue gases after the regenerator, which account for about 30% of the total energy consumption

---

[7] Barklage-Hilgefort, Hansjürgen, "Batch Preheating on Container Glass Furnaces," 69th conference on Glass Problems November 4-5, 2008, Columbus, Ohio. Page 133.

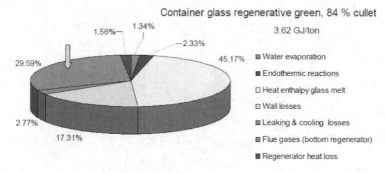

Fig. 1: energy flow of a typical container glass furnace [8]

The flue gas temperatures after the regenerator may vary, depending on furnace type, age, regenerators etc., between 370 and 600°C.

In case of oxy-fuel furnaces flue gas temperatures are even higher through the lack of regenerators. This energy is usually lost via the stack and constitutes the starting-point for batch- and cullet preheating.

HISTORY AND TECHNOLOGY OF BATCH- AND CULLET PREHEATING

Batch & Cullet Preheating is not new. Already decades ago, glass technology professionals saw the necessity to recover the waste heat. The advantage of batch & cullet preheating are twofold. On the one hand, a significant amount of energy can be saved (holding pull constant)- on the other hand, also the furnace pull can be increased in the same range as the energy consumption of the furnace decreases. When holding the pull constant, the furnace lifetime can be increased due to lower melting temperatures in the furnace, as some pre-reactions of the batch have been replaced into the preheater (such as water evaporation)[9]. Usually the optimum operating level will be a trade-off between these two parameters. Practical experiences have shown that the best results are obtained when also the melting rate is increased- thus batch preheating for future installations may be seen as an integral part of the whole production system.[10]

To make use of the waste gas, different concepts existed, such as using boiler systems, or usage of the heat for any other application, like heating buildings. Because of high investment costs and lower efficiency of boiler systems and limited application possibilities for direct heating of buildings etc., other concepts had to be found.

---

[8] Beerkens, Ruud, TNO. 1.October 2008 NCNG-Senter Novem-TNO workshop
69th Conference on Glass Problems Columbus OH 3. & 4. November 2008.
[9] As a result of the lower gas consumption, volume flows in the furnace can be reduced and such heat losses minimized.
[10] With pull increase, energy savings up to 20% are possible. Beerkens, Ruud, TNO. 1.October 2008 NCNG-Senter Novem-TNO workshop.
69th Conference on Glass Problems Columbus OH 3. & 4. November 2008.

Basically, there were two batch preheating systems which were both developed in the early 1980ies and were adopted by the market to some extent.[11] One is the so called Nienburger-type direct batch & cullet preheater, which was designed for applications with high cullet content (> 60%).

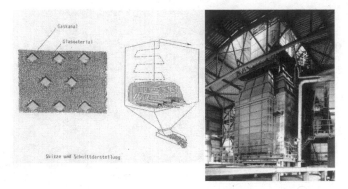

Figure 2: The Nienburger-type batch & cullet preheater

The basic concept of this design follows the direct-principle, meaning, there is a direct contact between flue gases and the batch. The flue gases are directed in a cross/counter-flow manner through so called roof-elements, which are open at the lower side and such create a hollow space under these roofs. The material is moving vertically downwards and is- while having direct contact with the gases- being preheated. The flue gas enters the preheater at the lower part with a temperature of 400–450°C and leaves the preheater in the upper part with a temperature of about 250-325°C.

As the gas channels are open at the bottom side, acid components in the flue gases HCI, HF, SOx and SeO$_2$ are partly absorbed by the earth-alkali compounds in the batch (soda, limestone, dolomite). Thus, the preheater partly also works as a scrubber.[12]

Due to the direct contact of batch and flue gases, dust emissions are increased significantly and thus the technology requires appropriate filter systems, such as large electrostatic precipitators or bag house filters. Measurements from the operator show an increase of dust concentration before the preheater from 96.5 mg/m^3 to 1675 mg/m^3 after the preheater. Behind the precipitator, a dust concentration of 22.4 mg/m^3 was measured- a value comparable to furnaces without batch preheating technology. Taking this into consideration, the use of an electrostatic precipitator becomes a must when applying

---

[11] There were also the PRAXAIR-Edmeston, and the SORG system. However these systems were designed for the application with very high cullet content of 85-90% and are therefore also called "cullet preheaters". Other patents were filed for example by OWENS-CORNING 1984, "Method of Preheating Glass Batch". This method utilizes a rotating drum for the heat transfer- at a commercial level this system has not been put through.

[12] Beerkens, Ruud, "Energy Saving Options for Glass Furnaces & Recovery of Heat From Their Flue Gases And Experiences With Batch & Cullet Pre-Heaters Applied In The Glass Industry," 69th conference on Glass Problems November 4-5, 2008, Columbus, Ohio. Page 153.

this technology. Latest modifications of this system aimed at reducing gas velocities in the preheater and reducing carry-over. Five of such systems have been built in the 80ies and 90ies, three of them are still running. The others stopped because of general glass production capacity reductions. The overall energetic results of the system show savings of about 15% with batch and cullet preheating at 300-325°C.[13]

THE LATEST INSTALLATION IN EUROPE
Lately- in 2010- a new installation of this type for a 350 mt/d container glass U-flame furnace in the Netherlands has successfully been installed and put into operation.[14]

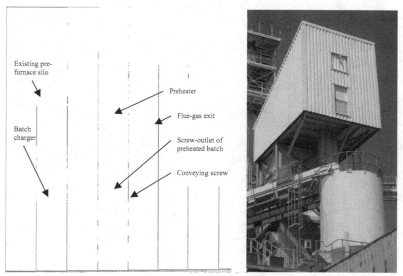

Figure 3: the latest batch preheating installation in Europe in 2010 (Copyright: Zippe Industrieanlagen GmbH)

A large european container glass producer decided to integrate a batch preheating system at an existing furnace. The new add-on-system starts at the existing pre-furnace-silo. Through a Y-section below the silo, the material is fed to a vibratory tray feeder that conveys the batch to the elevator and thus to the top of the preheater. Over another vibratory tray feeder the material is brought into the preheating system. The outlet of the system is guaranteed by in-line screw feeders that transport the batch to the collecting screw feeder. Finally, the batch is conveyed to the existing batch charger and brought into the furnace. In case of maintenance, a bypass is foreseen and the normal way through the pre-furnace silo is taken. The flue gases are taken from the existing waste-gas pipes though a bypass. In case of need, the whole system can be bypassed and the normal production process would be applied.

---

[13] Details of this systems are not be elaborated here- they might be found in Barklage-Hilgefort, Hansjürgen, "Batch Preheating on Container Glass Furnaces," 69th conference on Glass Problems November 4–5, 2008, Columbus, Ohio.
[14] System supplied by ZIPPE Industrieanlagen GmbH.

A few key facts characterize the system:

- furnace type: regenerative U-flame, 1 doghouse
- throughput: ~350 mt/d
- flue gas inlet temperature: max. 450°C
- flue gas outlet temperature: 220-230°C
- total weight: ~320 mt including steelwork and batch filling
- dimensions: 4700mm x 5400mm x 13000mm

The system was put into operation in July 2010 and first measurements show significant energy savings. As the last preheater-installation was implemented around 1996, this installation represents the latest one in 15 years.[15]

Another type is the so-called Zippe-type indirect batch preheater. Four of such systems have been built in the 1990s, the latest one was installed 1996 at a container glass furnace and has stopped its successful operation in 2010 because of a furnace shut-down.

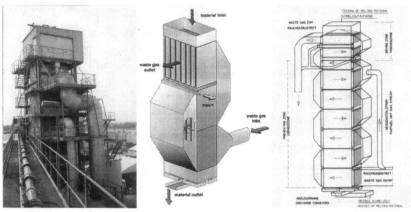

Figure 4: The Zippe-type Batch preheater (Copyright: Zippe Industrieanlagen GmbH)

The main difference with this system is that there is no direct contact between the batch and the flue gases. The preheater is being fed with batch in the upper inlet zone and the material is moving down by gravity with approximately 1–1.5 m/s. The waste gas enters the system at the bottom part and is being led upwards in a cross-counter flow and exits the system at the upper part with about 190–240°C. Devaporizing modules were designed and installed between the single modules to let the moisture exit the system. In the top section, the vapours have to be released without condensation and can be added to the hot flue gas.

---

[15] It is not known of any big-scale installation in Europe in the glass industry during that time.

Due to non-existing direct contact between batch and flue gas, there is no increase in dust concentration and no chemical reactions between substances of the flue gas and the batch can occur. Preheat temperatures also practically remain at a level of 250–325°C. In combination with an increased pull, energy savings of 15–20% have been found in the installation in the Netherlands. Also with this preheater, in case of usage of electric boosting, the need for boosting is lowered and the comparable more expensive electric energy can be saved- leading to a reduction of energy costs higher than the relative savings of energy.

The following table shows the results of such a preheater installed in 1996.

Table 2: results of indirect preheater in the Netherlands, installed 1996

Betriebserfahrungen: Scherben- und Gemengevorwärmung PLM Dongen
Practical Experience: Cullet and Batch Preheating at PLM Dongen

Rauchgaseintritt/Waste gas inlet:	480 °C
Rauchgasaustritt/Waste gas outlet:	270 °C
Schmelzguteintritt/Batch input:	15 °C
Schmelzgutaustritt/Batch outlet:	280 °C
Gemengedurchsatz/Batch throughput:	15.5 to/h
Scherbenanteil/Cullet ratio:	65%
Erdgaseinsparung/Saving of natural gas	7,8 %
Stromeinsparung/Saving of electric energy	62,2 %
Gesamte. Energieeinsparung/Total Energie savings	14%
Energiekosteneinsparung/Saving of energy costs:	27%
Erdgaseinsparung/Saving of natural gas:	4.390.000 m³/year

The batch, including 65% cullet, is being heated up to 280°C at a throughput of 15.5 mt/h while using a temperature delta between 480°C waste gas input and 270°C waste gas outlet temperature. Total energy savings remain at a level of 14%. As more costly electric energy could be saved as a result of reduced electric boosting, a total saving of energy costs of 27% was achieved.

So the question arises, why batch preheating systems have not been adopted widely in the glass industry- given its significant energy saving potential. One point surely is that it is a substantial capital investment- roughly between 1.2–1.8 mil. Euro (1.5–2.3 mil. $) for a complete system- that needs to be justified. Also, relatively cheap energy prices prolonged return on investments. Furthermore, some improvements in the design of such systems had to be done.

Shortcomings of former preheating devices were:

- chemical batch reactions due to water migration (condensation, evaporation)
- fatigue of preheater- material because of corrosion and exposure to high-temperature
- changed bulk behaviour of cold and preheated batch
- charging and junction of cold and hot cullet/batch and carry-over in the furnace
- odour nuisance through burning-off of organic compounds from the cullet
- maintenance requirements and poor accessibility of batch preheating aggregates
- restrictions in cullet content (min. 50%)
- restrictions in batch moisture

While other shortcomings of these first generation systems could be solved, especially the maintenance effort due to occasional clogging of material in the indirect batch preheater had to be reduced significantly. Due to the physical batch moisture of appr. 3–4%, and also the chemically bound water content of the soda, from about 104°C on, evaporation starts and the physical and also the chemically bound moisture has to leave the preheating system. As, in a completely indirect system, there is no contact between flue gas and batch material, and thus moisture cannot be taken out by the waste gas stream, it has to leave the system in another way. Since, in the 1st generation system, the de-vaporization units were still not optimal, another solution had to be found.

To overcome these shortcomings of existing systems, ZIPPE has initiated a new R&D project in 2006. Partly, these results are now available and will be presented below.

THE NEW DEVELOPMENT OF THE 2ND GENERATION "ADVANCED BATCH PREHEATER" SYSTEM

The new system constitutes a hybrid between indirect and direct preheating-systems.[16] In many inhouse-tests, combined with theoretical modelling, the chemical behaviour of soda at different temperatures had to be studied as an essential starting point.

When heating up batch, a key reaction that is taking place is the release of moisture during the different phases of the soda transformation.

---

[16] The indirect characterization however surely prevails.

Table 3: the different soda phases at different temperatures

		water (% by weight)	Transition temperature into lower phase (°C)
free of water	Na_2CO_3		
monohydrate	$Na_2CO_3 * H_2O$	14,5	> 107
heptahydrate	$Na_2CO_3 * 7H_2O$	54,3	> 35
dekahydrate	$Na_2CO_3 * 10H_2O$	62,9	> -2

Pure soda holds 62.9 weight percent of water. Usually, in the glass industry, calcined soda (Na_2CO_3) is used. However, this soda is highly hygroscopic and is being enriched- depending on the ambient temperature and humidity- with water and transformed to monohydrate ($Na_2CO_3*H_2O$)- or, in the worst case even into the heptahydrate. After exiting the batch mixer, it may surely be assumed that the soda is at least in the state of a monohydrate by this point of time- meaning that there is still a minimum of 14.5 weight percent of water enclosed.

This effect was known from theory before, however the magnitude of the practical relevance needed to be verified. In a small-scale preheater, batch with 50% cullet and 1.5% moisture was heated up by two burners with 600°C.

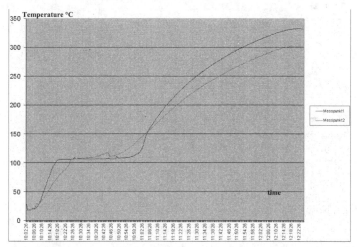

Graph 3: the effect of the soda transformation when heating up batch

Firstly, a rapid increase of batch temperature can be seen. However, even with the relatively low moisture content, at about 107°C, the effect of the water migration becomes obvious- the temperature remains at this level for a remarkable long time[17] before the transformation has taken place and the temperature is rising again. Also, considerable clogging occurred in the aggregate. So the goal for a new system was to find a system that is able to bring out the moisture from the system reliably, especially at the point of time when the main amount of water is being set free. Also, the first generation system had so-called de-vaporization modules- located every 1000mm between the preheating modules- that were designed to withdraw the condensate by suction into the waste gas pipe. However these modules, depending on batch moisture and cullet content, tended to clog and regular cleaning was necessary. Also, these modules guaranteed a de-vaporization only section-wise, thus a continuous process had to be found.

In the following a schematic drawing (cross-section) of a module of the new developed batch preheater (ABP[18]) is shown.

The flue gases enter the system at the lower part and stream upwards in a counter-flow manner while being directed by deflectors.

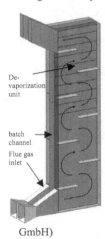

Especially in the drying section, a continuous devaporization needed to be guaranteed. Thus, especially laid-out slots were constructed that were able to let the moisture exit the system.[19] To improve the heat conductivity, the flue gas channels were fully integrated into the system. This system was completed with a pre-drying unit at the top and a mechanical anti-agglomeration device[20] for applications with cullet contents below 30 per cent. As in the previous system, the batch moves down vertically in the batch channels by gravity and heat-transfer is secured by convection through the channels.

De-
vaporization
unit

batch
channel

Flue gas
inlet

Figure 5: concept of the new ABP[21] (Copyright: Zippe Industrieanlagen GmbH)

Some characteristics show the differences between the first generation systems and the ABP:

Table 4: comparison between old and new batch preheater type ABP

	Old type	ABP
Working principle	cross-counter flow	cross-counter flow
Construction	modular	modular
Heating mode	indirect	Indirect/semi-indirect

---

[17] In this case for about 40 mins.
[18] ABP refers to the "Advanced Batch Preheater" Concept and will also be used in the following.
[19] Patent registered.
[20] Patent registered. Not shown on this drawing. This device mechanically loosens the batch in the pre-drying zone at the top of the preheater to avoid any sticking.
[21] Patent registered.

Waste gas conduction	internal / external	completely internal
Steam discharge	yes	yes, improved
Max. waste gas temperature	600 °C	650 °C
Max. batch humidity	max. 0.5%	tested > 5%
Min. cullet value	min. 60%	tested < 20%[22]

The new type is mainly based on the indirect system, however, with now small openings in the flue gas channels for removal of the evaporated water from the batch & cullet. It is a modular system that is adapted to the needed capacity. Also the cross counter flow was maintained as it guarantees favourable degree of efficiency. The main improvement and benefit is the applicability of this system with low cullet percentages.

PRACTICAL TESTS OF THE ABP
Even though a considerable level of experience in batch preheating was achieved through the last decades, a practical test on a glass melting furnace under real conditions had to be performed to prove the superiority and the practical safety of a new system. An in-house test, in whatever size would never reflect reality and would not give security and confidence high enough for future installations in glass furnaces under real conditions. For this, a cooperation with an European container glass producer was formed and also a leading furnace specialist and an international and reputable glass research institute were integrated into the project. As batch preheating technology must be seen as a system, it was regarded as valuable that different companies in a related sector would exchange know-how.

A long-term test of the new system at a container glass furnace was planned and performed. The project endured for about 2 years.

The basic data of the furnace were:

- tonnage of 320 to/d
- regenerative U-flame
- 2 doghouses
- 17% cullet addition
- 3% batch humidity
- 4450 MJ/ton actual energy consumption
- 8% electric boosting

Thus, it can be said that this shows an example of a typical European container glass furnace- however with a very low cullet percentage. So the conditions were a challenging test for the new system, as with existing systems, an operation with such low cullet content would have not been possible and would have resulted in severe clogging and caking problems that finally would have stopped the material flow and the whole system.

To guarantee a significance of the results, the test preheater was laid out for a capacity of 1/8 of the total tonnage, for about 40 tons per day.

The general layout is shown in the following:

---

[22] In a small scale preheater, tests with cullet contents below 10% were successfully conducted.

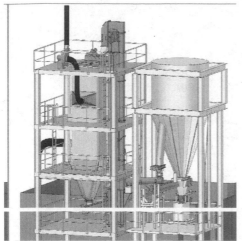

Figure 6: Layout of the test system (Copyright: Zippe Industrieanlagen GmbH)

The aggregate is installed next to the existing pre-furnace silo, where an additional outlet is foreseen. Through a vibratory tube feeder, the material is charged via an elevator to the top of the preheater. To guarantee a homogenous and even distribution into the preheater, a so-called anti-agglomeration device[23] is installed at the top of the system. The sink speed of the batch is approx. 1 m/h. At the outlet, a screw feeder conveys the preheated material to the existing batch charger and into the furnace.

The average flue gas inlet temperature was at about 375°C, the outlet temperature at about 230°C. So the available waste gas temperature was lower than typically, resulting in lower preheating temperatures. The material preheat temperature was at about 210°C. In case of higher flue gas inlet temperatures, higher preheating temperatures would have been achieved.[24]

RESULTS OF THE TEST SYSTEM
A major goal of the test was to verify that the system allows a reliable operation with low cullet contents, so the cullet content had to be reduced gradually. It was started with 70% cullet and 3.2% moisture. The furnace pull was about 305t/d and throughput of the preheater was about 36mt/d. A preheating temperature of above 200°C was achieved, which was lower than planned, due to the lower flue gas inlet temperature of about 360°C. Each test sequence was performed for a period between 1 and 3 days. The following table sums up the major results.

---

[23] Patents registered.
[24] The relation between flue gas inlet temperature and material outlet temperature is almost linear.

Table 5: main test results

70% cullet 3.2% moisture	Furnace pull 305t/d, preheater throughput 36t/d	No clogging, safe operation
50% cullet 3.2% moisture	Furnace pull 305t/d, preheater throughput 50t/d	No clogging, safe operation
40% cullet 3.2% moisture	Furnace pull 310t/d, preheater throughput 40t/d	Some Clogging in a few material chutes
Modification of surface area of the preheater, change in material charging and distribution		
70% cullet 3.2 moisture	Furnace pull 310t/d, preheater throughput 40t/d	No clogging, safe operation
50% cullet 2.5% moisture	Furnace pull 310t/d, preheater throughput 36t/d	No clogging, safe operation, higher dust concentration in flue gas
40% cullet 2.5% moisture	Furnace pull 310t/d, preheater throughput 36t/d	No clogging, safe operation, higher dust concentration in flue gas
30% cullet 2.5% moisture	Furnace pull 310t/d, preheater throughput 40t/d	No clogging, safe operation, higher dust concentration in flue gas
<20% cullet 2.5% moisture	Furnace pull 310t/d, preheater throughput 40t/d	No clogging, safe operation, higher dust concentration in flue gas
Modification to lower under-pressure in system and thus lower dust concentration		

The results of these tests have clearly shown that it is possible to preheat batch with cullet contents below 20% and maintain a safe operation. These tests were performed over a period of several months at an existing glass furnace in operation. These positive results represent a novelty for the glass industry. The next trials have the aim to lower the dust concentration in the flue gases and are planned for the fourth quarter in 2010. In case of the expected positive results, it can be said that a sustainable concept of batch preheating has been found for a wide range of applications.

IDEAL PRECONDITIONS FOR INSTALLATION OF A BATCH & CULLET PREHEATER
Although theoretically, a batch & cullet preheater can be installed at almost every furnace, there are factors that promote an installation. Typically, these devices are installed at container glass furnaces only.[25] These furnaces are run with a higher amount of cullet and also the batch charging area makes an integration of a preheater and also the charging of hot and dry batch easier. Often, green glass furnaces are run with high cullet percentages and also a high amount of electric energy which favours the benefits of a preheater. If the furnace is operated with electric boosting, even higher energy cost savings can be obtained through the substitution of the relatively more expensive electricity. When applying a preheater at a furnace with a high pull rate, relative investment costs will be lower and also the relative energy savings will be higher and thus the payback time shortened.

---

[25] At the point of time, the author has no knowledge of any preheater installed at a float-glass plant.

With the new ABP, also applications with low cullet percentages of about 20% and lower are possible according to the latest practical tests. These may enlarge the potential fields of operation significantly. Also, when taking into account an indirect system, large filters, or even additional filter equipment are not necessarily needed. Thus, lower investment cost may lead to better economic feasibilities.

Furthermore, there are of course constraints in space that have to be considered when integrating a preheater as an add-on.

SUMMARY AND FUTURE OUTLOOK
Modern batch preheating offers significant energy saving potentials and is suitable for many furnaces, and also applications with less cullet availability. Tests have shown that the latest generation is able to handle batch with less than 20% cullet.
Long-term experiences show a safe operation of preheaters. Rising energy prices and tightening regulatory constraints increase the necessity to invest in such systems. As about 30% of the melting energy is still lost through the flue gases, batch preheating represents the starting point with the highest potential.

Future improvements will have to focus on the eventual carry-over in the furnace. As all batch preheaters deliver very dry batch, the batch charging situation must eventually be adapted. Although existing systems run reliably with regular batch chargers, to decrease dust formation in the furnace, regenerators, and also nearby the doghouse, eventual modifications of the batch charging must be taken into consideration when installing a batch preheater. This especially for end-port fired furnaces, where the batch is almost directly exposed to the gas flows from the flames.

Taking into account the latest experiences and also economical and ecological frameworks surrounding the glass industry, preheating technology may be considered more than ever seriously for new green-field furnaces and existing furnaces with favourable conditions.

REFERENCES
1. Barklage-Hilgefort, Hansjürgen, "Batch Preheating on Container Glass Furnaces," 69th conference on Glass Problems November 4-5, 2008, Columbus, Ohio. Page 133.
2. Beerkens, Ruud, TNO. 1.October 2008 NCNG-Senter Novem-TNO workshop 69th Conference on Glass Problems Columbus OH 3. & 4. November 2008.
3. Beerkens, Ruud, TNO. 1.October 2008 NCNG-Senter Novem-TNO workshop.
4. Beerkens, Ruud, "Energy Saving Options for Glass Furnaces & Recovery of Heat From Their Flue Gases And Experiences With Batch & Cullet Pre-Heaters Applied In The Glass Industry," 69th conference on Glass Problems November 4-5, 2008, Columbus, Ohio. Page 153.
5. Conradt Reinhard (2003)- HVG-Mitteilungen 2037, 1-6 (2003)
6. EUROSTAT: The european department of statistics (Brussels). http://epp.eurostat.ec.europa.eu/portal/page/portal/eurostat/home/
7. Glüsing, Ann-Kathrin, "Preheating Devices for Future Glass Making, a 2nd generation." 67th Conference on Glass Problems, Columbus Ohio, 30th Oct.- 1st Nov.
8. Neckermann Gerhard, Wessels Hans. Die Glasindustrie-ein Branchenbild. Deutsches Institut für Wirtschaftsforschung. Heft 95, 1987

OXY-FUEL CONVERSION REDUCES FUEL CONSUMPTION IN FIBERGLASS MELTING

John Rossi
Fiber Glass Industries, Inc.

Michael Habel, Kevin Lievre, Xiaoyi He, and Matthew Watson
Air Products and Chemicals, Inc.

ABSTRACT:
In recent years, oxy-fuel combustion has become widely embraced for the many benefits it brings to glass melting operations. Air Products has facilitated acceptance of this change to the glassmaker's process with a sustained investment in people, R&D facilities, and activities that continuously brings new and innovative products to the industry. The Cleanfire® HRi™ high radiation burner is one of the latest examples of this progress. In just a few years since its commercial release, the glass industry has made it the preferred technology when oxy-fuel burners are used for boosting an air-fuel furnace, or to convert the entire furnace to oxy-fuel firing.

In addition to installation at the time of scheduled furnace repairs, in most instances this technology can be installed on-the-fly while the furnace is in production. This paper will provide review of a full-furnace conversion accomplished at furnace rebuild in the fiberglass industry. Since the glass company's primary motivation to convert their melting operation to oxy-fuel was fuel savings, most of the focus will be about the fuel efficiency improvement. The review will include discussion of burner features that act to maximize combustion efficiency by increasing the amount of energy transferred directly from the flame to the batch materials and to the molten glass. A comparison of operational data before and after conversion from air-fuel to oxy-fuel combustion as well as operator observations is included.

INTRODUCTION:
In recent years, oxy-fuel combustion has become widely embraced for the many benefits it brings to glass melting operations. Air Products has facilitated acceptance of this change to the glassmaker's process with a sustained investment in people, R&D facilities, and activities that continuously brings new and innovative products to the industry. The Cleanfire® HRi™ high radiation burner is one of the latest examples of this progress. In just a few years since its commercial release, the glass industry has made it the preferred technology when oxy-fuel burners are used for boosting an air-fuel furnace, or to convert the entire furnace to oxy-fuel firing. This paper will describe the recent conversion of a recuperative furnace at Fiber Glass Industries.

Fiber Glass Industries (FGI) is a leading supplier of specialty fiberglass products, located in Amsterdam, NY. To differentiate from their competitors, they are entering new markets with more products, which leads to highly variable production demands The production facility consists of a single furnace operation, and the company has had limited experience with oxygen in its process. FGI was planning for their next furnace rebuild and was evaluating alternatives that would allow them to reduce their energy consumption, improve product quality, and provide the production flexibility required to serve their customers.

The melting operation at FGI used a recuperative furnace with 30 air-fuel burners. Each burner was adjusted manually using a local differential pressure reading. The main bottleneck to increasing production was the furnace, and because of the space constraints imposed by the surrounding building, it was not possible to increase the footprint of the furnace during the next rebuild. Typically the furnace was operating at its maximum capacity, and therefore even small process upsets had the potential to cause significant production disruptions to downstream processes. As a single furnace operation even the smallest disruption could jeopardize customer deliveries.

During their next cold repair, which took place in the spring of 2007, FGI wanted the ability to increase their production flexibility, especially with the ability to expand their furnace output quickly. They also wanted to improve their furnace efficiency to hedge against uncertain natural gas prices, which at the time of the scheduled cold repair were in the range of $10-12/MMBTU. In addition, FGI wanted to improve the overall bushing efficiency in an effort to reduce their cost per pound of product, and as such furnace stability plays a major role. Finally, FGI wanted to be proactive about improving their emissions performance and wanted to incorporate current best available technology.

Given all of FGI's requirements, Air Products and Chemicals, Inc. (APCI) recommended converting their furnace to a full oxy-fuel operation. Unlike air-fuel combustion, which as the following approximate stoichiometry:

$$CH_4 + 2O_2 + 7.5N_2 \rightarrow 2H_2O + CO_2 + 7.5N_2 (+ xNO + yNO_2 + zN_2O) + \ldots,$$

nitrogen is nearly eliminated from the combustion space, thereby substantially reducing the production of oxides of nitrogen (NOx). Oxy-fuel combustion has the following approximate stoichiometry:

$$CH_4 + 2O_2 \rightarrow 2H_2O + CO_2,$$

which results in a large reduction in flue gas volume. This is shown in Figure 1.

Oxy-Fuel Combustion

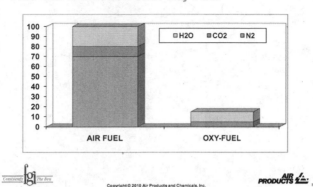

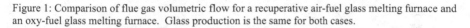

Figure 1: Comparison of flue gas volumetric flow for a recuperative air-fuel glass melting furnace and an oxy-fuel glass melting furnace. Glass production is the same for both cases.

This large reduction of flue gas volume is because N2 is no longer available to act as a ballast to carry out energy with the exhaust. For oxy-fuel combustion, this eliminates the need for heat recovery equipment. It is well established in the glass industry[1,2] that oxy-fuel combustion results in a more efficient melting operation when compared to pre-heated air-fuel combustion and this leads to reduced fuel consumption per ton of glass produced, as shown in Figure 2.

[1]Eleazer, P. B. and Slavejkov, A. G., "Clean firing of a glass furnace through the use of oxygen," in *Proceedings from 54th Conference on Glass Problems*, The American Ceramic Society, Westerville, OH, pp159-174, Oct. 1993.
[2]Beerkens, R. "Energy balance of glass furnaces: Parameters determining energy consumption of glass melt process," in *Proceedings from 67th Conference on Glass Problems*, The American Ceramic Society, Westerville, OH, pp103-116, Oct-Nov. 2006.

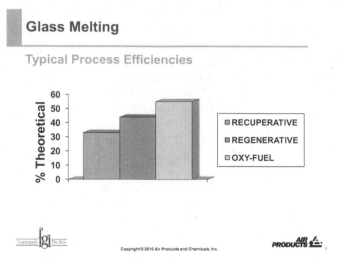

Figure 2: Theoretical efficiency of recuperative, regenerative and oxy-fuel furnaces.

In addition, since the flame is substantially hotter, oxy-fuel combustion allows for glazing over the batch pile to occur sooner. Fast glazing coupled with the lower velocity of the fuel and oxygen entering the furnace results in lower particulate emissions. The decreased flue gas volumes and lower flame velocities also results in a reduction of the volatilization of the more volatile glass species. This reduction of lost batch components leads directly to a more homogeneous melt and higher glass quality.

To summarize, oxy-fuel combustion can provide the following benefits:

- Smaller waste gas handling system due to decreased volume of flue gases

- Increased production as a result of increased flame temperature leading to the ability to intensify the production rate from a given furnace foot-print

- Increased energy efficiency due to the increased available heat from oxy-fuel combustion

- Operational flexibility due to the ability to increase and decrease production from the furnace

- Reduced pollutants, such as NOx and particulates, being emitted from the furnace

- Improved furnace stability and higher product quality.

THE OXY-FUEL SOLUTION
As a part of the overall conversion of FGI's furnace to oxy-fuel APCI provided the services and equipment to minimize any disruption to FGI's overall operation. Air Products provided a turnkey

installation that included the Liquid Oxygen storage and delivery system, safety and flow distribution equipment, supply piping, Cleanfire® HRi™ burners, safety training and operational support. Computational fluid dynamics (CFD) modeling was carried out to determine burner locations, flue sizing and operating conditions; this will be described more in the next section.

A liquid oxygen supply system is installed to allow for the safe delivery of liquid oxygen (LOX) to the storage tank. Road tanker delivered, liquid oxygen is stored in a cryogenic storage vessel at the glass factory. The heavily insulated cryogenic storage tank, holds the oxygen in the liquid state at about -300oF (-186oC). The oxygen is typically vaporized using ambient air vaporizers and typical storage pressures range up to about 350 psig (24 Barg).

The oxygen flow control skid is required to receive vaporized oxygen from the LOX storage tank and safely deliver it to the burner. Typically this is coupled with a fuel flow control skid to form a complete combustion control system. The function of the combustion control system is to safely deliver, precise amounts of fuel and oxygen to the oxy-fuel burners. The fuel flow rate is set by the operator and the oxygen flow rate is determined by the stoichiometric oxygen requirement of the fuel. The oxygen flow is metered in a precise and often automatic ratio to the fuel flow. With an automatic system the control system continuously monitors supply parameters, to ensure that if unsafe conditions occur, the system alarms and achieves a safe state. Typically, pressure and temperature corrected mass flow is measured and controlled for both the fuel and oxidant, and a separate control loop is used for each burner. Figure 3 below depicts a typical control system for use with oxygen and natural gas. Appropriate distribution pipework must be installed from the oxygen and fuel sources to each burner.

Figure 3: Typical control system for use with natural gas and oxygen.

The back-of-burner safety hardware includes local, manually operated valves, to ensure that fuel and oxygen can be locally shut off during burner installation and removal. A set of non-return valves (also commonly called check valves) are installed to prevent the risk of either fuel or oxidant travelling back into the other fluid's distribution pipework. Finally a set of flexible hoses are installed. For safety reasons, these hoses are as short as possible, suitable for the maximum allowable pressure, suitable for high temperature, and mounted to avoid distortion, whiplash or accidental damage. In the case of oxygen components, all materials are selected to be compatible, safe for use with oxygen, cleaned for oxygen service, and installed with special precautions followed for equipment to be used in oxygen service.

Flat flame, oxygen staged Cleanfire® HRi™ burners were provided for FGI's conversion to oxy-fuel. A brief overview of the combustion theory behind the burner is given below, and more detail can be found elsewhere[3,4,5]. The burner is available in multiple sizes to suit the various segments of the glass industry and glass producer's needs. For FGI's furnace, a size Mini burner was chosen which is capable of firing between 0.25 and 4.0 MMBTU/hr.

As part of the commissioning package Air Products provided safety and operational training to ensure FGI was comfortable with the operation of their oxy-fuel equipment.

Throughout the conversion, APCI technical experts were onsite to provide process expertise to help FGI step through the conversion with minimal disruptions.

BURNER SELECTION

Based on the FGI's requirements and our previous experiences, Air Products recommended the use of eight Cleanfire® HRi™ Mini gas burners. The size Mini burner has been used in a number of glass industry segments, ranging from container, frit, mini-float and fiberglass furnaces. It is typically used in smaller furnaces, and in situations where the thermal power range between 0.25 and 4 MMBTU/hr, and flame lengths ranging between approximately 1 – 5 ft are required.

The burner construction consists of three main components (Figure 4). First, on the left of the figure is the burner which directs fuel to a flat flame nozzle and oxygen to the annular space surrounding the fuel nozzle and to a separate staging port beneath the fuel nozzle. While the overall flow of oxygen to the burner is controlled by the flow control skid, the proportion of oxygen directed to the staging port can be adjusted by using the staging valve. Second, the burner mounting bracket forms an interface between the burner and the refractory burner block. The mounting bracket is bolted to the burner block, while the burner is connected to the bracket via for quick-release clips for easy removal. The mounting bracket helps to align the fuel and oxidant nozzles inside of the burner block. Appropriate gasketing is used to ensure a tight seal. Third, the refractory burner block consists of an upper opening, the precombustor, and a lower opening, the staging port.

[3] Slavejkov A. G. et al., U.S. Patent No. 5,575,637. Washington, D.C., U.S. Patent and Trade Office.
[4] Slavejkov A. G. et al., U.S. Patent No. 5,611,682. Washington, D.C., U.S. Patent and Trade Office.
[5] D'Agostini M. D., U.S. Patent No. 7,390,189. Washington, D.C., U.S. Patent and Trade Office.

Cleanfire® HR*i*™ Burner Technology

Air Products' Patented High Radiation Burner

Copyright © 2010 Air Products and Chemicals, Inc.

Figure 4: Cleanfire® HRi™ Burner Technology

The basic principle of the Cleanfire® HRi™ burner is that the flame starts at the metal burner nozzle and propagates through the refractory precombustor without touching the precombustor's walls. Oxygen always flows around the flame to protect the burner block from high flame temperatures. The oxygen staging function of the burner allows some of the combustion oxygen to be diverted through the staging port just beneath the primary precombustor. There are multiple benefits to staging – separating a portion of the oxygen required for combustion oxygen:

- Staging the oxygen away from the flame delays combustion which has the effect of elongating the flame. This leads to better flame coverage over the glass melt.

- Staging the oxygen away from the flame substantially reduces the peak flame temperature. When tramp air or fuel bound nitrogen is present, this has the effect of reducing NOx.

- Staging the oxygen away from the flame lowers overall flame momentum allowing for adjustability based on the degree of local turbulence for a given burner position.

- Staging the oxygen underneath of the flame results in a flame structure with an optically thick sooty or smokey topside and while the staged oxygen reacts with unburned fuel on the underside of the flame highly producing a highly radiant and luminous underside. This overall effect is directional radiation whereby the bright underside of the flame is preferentially directing thermal radiation downward toward the glass tank for high efficiency glass melting, while the radiation upward toward the crown is partially occluded by the sooty top layer.

As the staging valve is opened more, the overall effect is enhanced by increased levels of staged oxygen: longer flame, lower NOx, lower momentum, and increased thermal radiation directed downward to the glass melt, as shown in Figure 5.

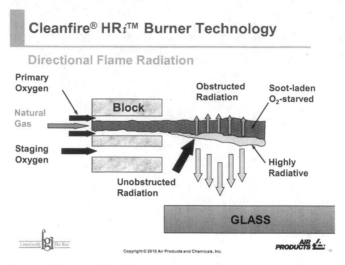

Figure 5: Effect of oxygen staging on direction radiation.

CFD MODELING

FGI's recuperative air-fuel furnace was 9.75 ft wide by 33 ft long and had 30 air-fuel burners. The air-fuel combustion operation was modeled using computational fluid dynamics (CFD) to obtain a base-line operating point against which predicted oxy-fuel operations could be made. Commercially available software Fluent (ANSYS INC, 2006) was used for this purpose. This also allowed the air-fuel operating measurements to be checked against CFD model-predicted results to check the accuracy of the CFD model. Figure 6 shows the combustion space temperature distribution within FGI's original air-fuel furnace. Glass flow and exhaust gas flow are indicated on the figure. The temperature distribution is shown in the plane of the burners, and the peek temperature in this plane is about 3500°F. As shown in Figure 7, the simulation faithfully revealed the temperature distribution inside the original air-fuel furnace.

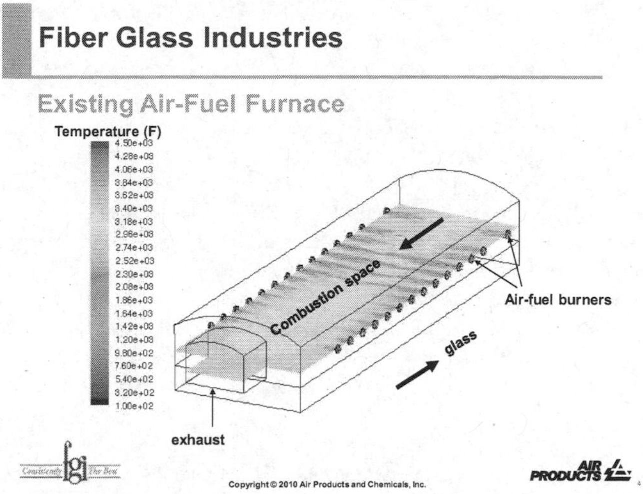

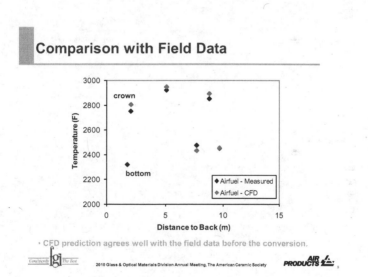

Figure 6: Combustion space temperature distribution within FGI's original air-fuel furnace.

Figure 7: Comparison of measured and CFD-predicted crown and glass (bottom) temperature distribution in the furnace for original air-fuel operation.

After the air-fuel operation benchmark study, we carried out a parametric study to identify the optimal design for the full oxygen furnace. In this project, the burner location is pretty much fixed because of restrictions on the furnace structure. Consequently, the optimization was done via adjusting the firing profile. The total firing rate was set at 55% of the original air-fuel case based on previous experiences. The first set of firing profile was derived from the existing air-fuel combustion. The simulated temperature distribution based on such a firing profile was then compared to that in the air-fuel case. It was found the temperature was relatively low near the charging zone and relatively high downstream (Figure 8, case study "oxy-fuel 1"). Subsequently, the firing profile was adjusted via moving more firing power from the middle section toward the charging zone (Figure 8, case study "oxy-fuel 2"). The process was iterated several times until the temperature distribution became comparable with that in the air-fuel case (Figure 8, case study "oxy-fuel 3"). The final result confirmed that a 40%~45% fuel saving can be achieved by converting to Air Products' full oxygen-fuel combustion technology, while still maintaining the glass exit temperature.

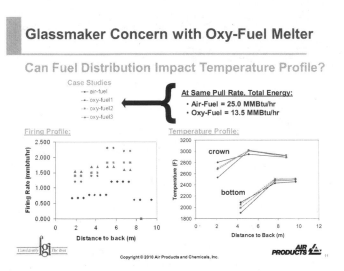

Figure 8: Effect of fuel distribution on furnace temperature profile.

Figure 9 showed the temperature inside the furnace after the conversion. Overall, the temperature distribution was very close to the temperature distribution in the original furnace. No overheating or under-heating was observed. As a final check of the validity of the CFD modeling exercise, the measured furnace temperature distribution was compared with the model-predicted furnace temperature distribution after the installation was completed. Figure 10 shows that there is good agreement between the model and the measured temperature distribution after the conversion to oxy-fuel operation.

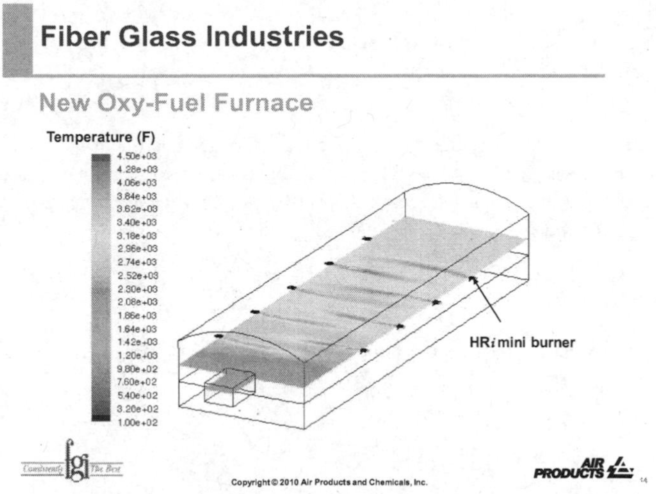

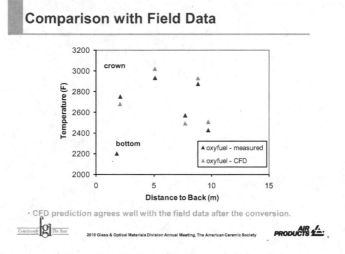

Figure 9: Combustion space temperature distribution within FGI's new oxy-fuel furnace.

Figure 10: Comparison of measured and CFD-predicted crown and glass (bottom) temperature distribution in the furnace after conversion to oxy-fuel operation.

Beside the design optimization, the simulation was also used to study the impacts of the operational variations. For example, it was found a 20% increase in total firing power is required for a 40% increase in the pull rate. On the other hand, the firing power needs to reduce by 20% if the pull rate decreases by 40%.

The size of the exhaust port was another subject of interests. In general, the exhaust port needs be sized down in a full conversion to maintain the proper furnace pressure. It is always a concern how the exhaust port reduction would impact the furnace operation. In this typical conversion, the temperature inside the furnace showed little dependence on the size deduction from 50% to 78% (Figure 11).

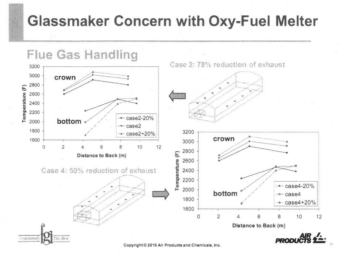

Figure 11: Impact of exhaust port on furnace operation.

SUMMARY AND CONCLUSIONS

The conversion of FGI's fiber-glass furnace from air fuel to oxy-fuel combustion was a great success. The post-conversion plant data showed a 50% fuel saving, which is consistent with the modeling prediction of 45% of fuel saving.

Simulation was instrumental for finding the optimal burner location and firing rate profile and was helpful to study impact of operation variations (pull rate) and furnace design changes (flue size). In addition fuel usage predictions, there was good agreement between the CFD predictions and field data of the temperature profile in both the crown and bottom of the furnace.

FGI's adoption of oxy-fuel combustion has lead to a number of improvements in their furnace operation, including furnace stability, enhanced profile control. In addition, FGI measured an increased yield through a significant bushing efficiency improvement. The furnace operation has been simplified – there are now 9 oxy-fuel burners and blocks, each individually controlled, versus the original 30 air-fuel burners and blocks with the operational and maintenance issues of using a recuperator and combustion air blower.

SOLAR GLASS MELTING

Matthias Lindig
Nikolaus Sorg GmbH, Lohr, Germany

SUMMARY

Photovoltaic glass is a special glass with integrated solar cells, used to convert solar energy into electricity. The production of the flat glass used for that purpose presents special challenges regarding melting technology. One of the challenges is the significant low content of iron oxide (<100mg/kg), which affects the primary and secondary refining conditions, the chemistry of this primary and secondary refining, the radiation heat conductivity and the glass melt flow pattern in the melter tank. Specific conditions regarding production of that special glass in conventional glass melting furnaces will be discussed in detail.

PRODUCT DESCRIPTION AND REQUIREMENTS

The solar cells are embedded between two glass panels and a special resin is filled between the panels, securely enclosing the solar cells on all sides. The sheet glass acts as a substrate as well as a protector against the ambient influence. The photovoltaic cell converts sunlight directly into electric current. The photon energy of the sunlight incidence produces positive and negative charges in the semi conducting layer between the two glass sheets and fed to the connecting electrodes. Each individual cell has two electrical connections, which are linked to other cells in the module to form a system which generates a direct electrical current.

The semi conducting substance can consist of mono crystalline, multi crystalline or amorphous silicon, so-called thin film cells. Other substances such as gallium arsenic cells with higher performance are in use in space technology.

Some special demands are put on the cover sheet glass which relate to performance and lifetime. These requirements are met by special adjustments in the glass composition, starting with the base composition which does not differ much from automobile or architectural flat glass compositions. The adjustments are comparatively small, but nevertheless they have a significant impact on the melting and refining conditions. These changes in conditions have to be compensated by adjustments to the conventional melting technology.

In addition to the chemical durability and the mechanical strength the light transparency is one of the key properties of the glass. The transparency of the glass is affected by the colouring ions of, for example, iron, chromium, nickel and cobalt. The semi conductive layer uses only a part of the total sunlight radiation spectrum for power generation. The important radiation transmission in the demanded wavelength range will be influenced by the iron oxide ions in the edge regions.

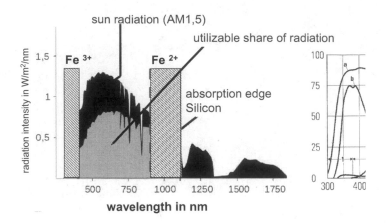

Figure 1: sun radiation vs. wavelength and utilizable share used by the semi conductor; the wavelength section affected by iron oxide is marked

In figure 1 it can be seen that the ferric oxide (3^+) affects the efficiency of the semi conductor less that the ferrous oxide (2^+) [1]. The iron oxide used to be present as an impurity in the silica sand. Raw materials with the lowest possible level of iron impurity are needed for the production of the photovoltaic sheet glasses. It has to be emphasized that the remaining low amount of iron oxide in the glass has a relative small influence on the overall semi conductor performance. The efficiency impact of the warming of the cell exposed to the sunlight radiation is an order of magnitude higher in comparison to the transmission reduction by the iron content. The nominal performance of the cell used to be related to the operating temperature of 25°C which is not the case in reality.

MELTING AND REFINING OF SOLAR GLASS- PRIMARY AND SECONDARY REFINING

The low iron oxide content in the melt has a significant influence on the chemistry of the primary and secondary refining and on the radiation heat transfer into glass. Standard soda lime silica glasses have an iron content of about 0.1 to 1% by weight and more looking at special glasses for the automobile industry. In the case of solar glass an iron oxide content of less than 100ppm is requested. The basic composition belongs to those glasses which are melted using sulphur (supplied as sulphate most of the time) as a fining agent. The chemistry of the sulphur refining is well known [2,3]. The sulphate added as sodium sulphate decomposes at high temperature into sodium oxide, sulphur dioxide and oxides. It has to be kept in mind that the iron (in ferrous state) acts in that case as a reducing agent which supports the production of sulphides or enhances sulphate decomposition at lower temperatures (reaction between sulphides and sulphate). There are two ways iron oxides (in ferrous state) can enhance sulphate decomposition at lower temperature.

Direct reaction between ferrous iron and sulphate, is producing ferric iron and SO_2 gases. Reactions at low temperatures between ferrous iron and sulphate results in generation of sulphide. This sulphide can react with sulphate at higher temperature (sulphate- sulphide refining) to form the fining gas SO_2.

Ferrous iron or other reducing agents promote the decomposition of the sulphates, because it absorbs the oxygen from the sulphate, forming SO_2 gas for the high temperature primary refining process:

$$Na_2SO_4 + 2Fe^{2+} \rightarrow 2Fe^{3+} + Na_2O + SO_2 \text{ (gas)} + O^{2-}$$

Reduced availability of this reducing agent, the ferrous iron, obviously has an impact
By implication reduction of ferrous iron in the batch can affect the efficiency of refining conditions in the melt. And this is what can be detected in practice.

Due to this, the addition of small amounts of antimony has become part of the refining operation for solar glass. The addition of antimony oxide should not be compared with traditional antimony refining. In the special case of a low iron soda lime glass melt the antimony oxide adopts the position of the reducing agent in order to support the sulphate refining. This is why the addition of nitrate, which is normally added to the traditional Antimony refining process, is not necessary. Test results with the addition of antimony and with or without nitrate addition confirm that theoretical approach. Also the second stage of refining will be affected by the addition of the Antimony. The secondary refining is influenced by the presence of other polyvalent ions in the melt. Seeds, rich in SO_2, can hardly be reabsorbed by the melt during controlled cooling of the melt in the case that no polyvalent ions are present. In case of ferric ions the following equation can be applied:

$$SO_2 \text{ (gas)} + 2O^{2-} + 2Fe^{3+} \rightarrow SO_4^{2-} + 2Fe^{2+}$$

During cooling of the melt the SO_2 seeds are reabsorbed as sulphate.

The second significant impact of the low iron content is visible on the heat radiation conductivity of the melt. The strong increase in the effective radiation heat conductivity is visible by the high glass melt temperature at the bottom of the melter.

The heat transfer in the glass melt consists of two terms. One is the heat transfer by radiation [4]. The second term is the heat transfer by convection of the melt. The resulting heat conductivity can be calculated using a couple of indices. The equations are given in figure 2.

$$\lambda_{melt} = \lambda_{rad} + \lambda_{konv}$$

$$\lambda_{conv} = \lambda^0 \cdot Nu$$

$$Nu = 0{,}42 \cdot (Gr \cdot Pr)^{0{,}25} \cdot \left(\frac{h}{L}\right)^{-0{,}25}$$

λ_{melt} heat conductivity glas melt
λ_{rad} radiation heat conductivity
λ_{konv} convective heat conductivity
Nu Nusselt number
Gr Grashoff number
Pr Prantl number
h depth of liquid
L horizontal expansion of liquid

Figure 2: Definition of the glass melt heat conductivity (radiation and convection)

Figure 3 shows a comparison of the radiative heat conductivity for green glass and low iron solar glass. The difference is in the range of two orders of magnitude. One might think that the buoyancy force will be much less in case of the low iron solar glass due to the higher bottom temperature in the melt, which should involve lower convective currents and finally lower convective heat conductivity.

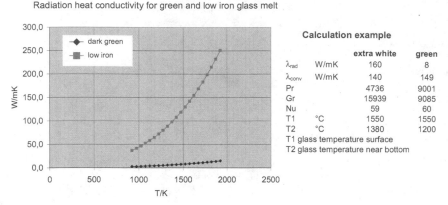

Radiation heat conductivity for green and low iron glass melt

Calculation example

		extra white	green
λ_{rad}	W/mK	160	8
λ_{conv}	W/mK	140	149
Pr		4736	9001
Gr		15939	9085
Nu		59	60
T1	°C	1550	1550
T2	°C	1380	1200

T1 glass temperature surface
T2 glass temperature near bottom

Figure 3: Extra white and green radiative heat conductivity and calculation example for both types of glasses (source GlassService)

The calculation of the situation for green and solar glass given in figure 3 shows that the strong reduction of the viscosity in the case of solar glass helps to compensate the reduction in buoyancy force. The Grashof number for extra white glass is even much higher than for the green glass case. This explains the fast response of temperatures in the melt on changes in fuel input. Mathematical modelling confirms these calculations and observations.

TYPICAL MELTING PROCESSING FOR SOLAR GLASS AND ALTERNATIVES

In addition to the all physical demands on the glass quality of photovoltaic technology and the resulting special melting conditions it has to be emphasized that the optical demands are also comparable high. For this reason extra fining agent is needed like Antimony oxide and special furnace design and operation pattern are requested. The optical quality demands are usually given by the DIN EN 572-2, a quality standard which was created for architectural glass. The high quality standard is not necessarily needed from the technical point of view, but nevertheless the high quality demand has become standard for solar glass panels.

In conventional glass melting furnaces glass quality, which in this case means the number of seeds, used to be influenced mainly by specific pull or residence time in the melter. Two different solar glass panel production types are in use at present.

FLOAT PROCESS

A smaller portion of the entire solar glass panel production is made using the standard float process, but the processing is relatively expensive. Existing float lines are usually used for this purpose without any special modifications. The critical feature of such standard melters is the comparatively shallow glass depth of about 1100mm to 1300mm. As mentioned above the bottom temperature will rise up to a critical limit when a solar glass composition is produced in such a tank. Due to this it is very common to significantly reduce the throughput in such a case. The disadvantage of the use of that manufacturing process in general is the panel surface resulting from that float process. Investigations have shown that a textured panel surface is more desirable, as it helps to disperse the sun radiation beam into the glass panel which leads to higher sunlight absorption and efficiency.

ROLLED PLATE GLASS PROCESS

Due to this the major part of the world solar glass panel production is made as rolled plate glass. The throughput of the existing furnaces for rolled plate glass varies. Usually a single line runs with about 130 metric t/d. For a single production line a regenerative end-port furnace is suitable, whereas more than one line would require a cross-fired furnace.

Those furnaces specially built for solar glass production should have a sufficient glass depth of about 1300 to 1500mm. The constraints of the refining process were mentioned above, and to account for these special design features may be provided inside the tank which support the flow pattern and the refining processes. One of these features might be a bubbling or barrier boosting system.

Both features might not be able to prevent poorly refined glass flowing rapidly forwards along on the bottom through the outlet of the tank, resulting in poor glass quality at the end of the line. A more efficient way to improve the refining conditions would be the use of a barrier wall or refining shelf. A couple of mathematical modelling studies of the performance of a melter with or without barriers were carried out by Sorg. All of these studies came up with the result that the refining shelf always improves the melting and refining conditions and also help to reduce the melter bottom temperature and might also to allow temperature reduction in the entire furnace.

The use of oxy-fuel heating was considered in several recent solar glass melter projects. One should keep in mind that the oxy-fuel firing in general increases the risk of producing more foam in the case of soda lime silica glass compared with standard gas-air fired furnaces. The problem is linked to the increased concentration of water in the combustion atmosphere. The water vapour migrates into the melt and from there into the existing seed. Partial pressure conditions of the gases dissolved in the melt and present in the seeds help to inflate these which results in the generation more gases from the melt and of a stable foam layer. Modifications to the batch composition and special adjustments to the furnace operation pattern will be necessary.

CONCLUSION

The very minor modification (lower iron content and sometimes more oxidized conditions) of the glass composition for producing solar glass panels results in very significant changes in melting and refining conditions. The rise in bottom temperature requires increased depth in the melter in order to prevent early bottom refractory damage and to increase the residence time of the glass.

A refining shelf is found to improve the refining conditions.

The refining chemistry (primary and secondary refining) is nowadays completed by the use of antimony oxide, whereby this application is not comparable with the standard conditions of nitrate antimony oxide very oxidized refining.

The theoretical consideration of the influence of the low iron content on the effective heat conductivity of the melt helps to understand that the flow speed is not reduced with low iron glass.

The flow speed will be even higher compared to a standard soda lime silica glass melt. From this point of view special attention should be addressed on the refining conditions in general.

REFERENCES:

1. H.Müller Simon
   Bedeutung des Eisengehaltes in Deckgläsern für Photovoltaik- Module
   Proceedings 83rd DGG Congress, Amberg, Germany

2. R.Beerkens, K.Kahl
   Chemistry of Sulphur in Soda-Lime-Silica- Glass Melts
   Phys.Chem.Glasses 43(2002) 4, 189-198

3. H.Müller-Simon
   Elektronenaustausch zwischen Paaren polyvalenter Elemente in techn.Gläsern
   Verlag Deutsche Glastechn.Gesellschaft 2007

4. A.Lankhorst, A.J.Faber
   Spectral Radiation Model for Simulation of Heat Transfer in Glass Melts
   Glass Techn.Eur.J.Glass Sci.Technol. April 2008, 49(2), 73-82

INTEGRATED AIR QUALITY CONTROL SYSTEM FOR FLOAT GLASS FURNACE

Michael Cheng
Fabrication Manager
Guardian Industries Corp., Kingsburg, California

Nathan Blanton
Sales Manager
GEA Bischoff, Inc., Memphis, Tennessee

ABSTRACT

In 2008 Guardian Industries Corp. conducted their second cold tank repair on their Kingsburg, California float glass plant. Changes to the operations included installation of Lowest Achievable Emission Rate (LAER) air quality control (AQC) technology. New regulations for glass furnaces adopted by the San Joaquin Valley Unified Air Pollution Control District required the replacement of the existing electrostatic precipitator for particulate control with a new and modern Air Quality Control (AQC) system for reduction of sulfur oxides (SOx), particulate matter (PM), and nitrogen oxides (NOx). The AQC system was commissioned mid-2008 and is successfully meeting initial permit emission limits.

First stage – dry scrubbing SOx with Trona. Furnace exhaust gases in excess of the normal operating temperature range are cooled by water injection prior to the first gas cleaning stage, where SOx and condensable particulates are removed. Trona is introduced into the duct prior to the scrubber as a dry sorption reagent.

Second stage – particle collection. An electrostatic precipitator (EP) removes filterable particulate matter. The captured EP dust is pneumatically transported to an intermediate storage silo for potential reintroduction into the batch. When the EP dust meets quality requirements, it can be recycled, thus eliminating the need to dispose as solid waste.

Third stage – NOx reduction. In the final gas cleaning stage, furnace exhaust gas NOx emissions are reduced in the Selective Catalytic Reduction (SCR) unit. Aqua Ammonia is injected upstream of the SCR and NOx is reduced as it reacts with the aqua ammonia on the surface area of the catalyst bed.

The paper describes the AQC plant supplied by GEA Bischoff, and performance data are given. It should be noted that the NOx reduction stage is the first application in the US.

BACKGROUND

After 30 years of production, Guardian Industries Corp. (Guardian) ceased operations to conduct their second cold tank repair.

Planned improvements to the operations during the repair required a New Source Review and installation of Lowest Achievable Emission Rate (LAER) pollution control technology. New regulations for glass furnaces adopted by the San Joaquin Valley Unified Air Pollution Control District required the replacement of the existing Electrostatic Precipitator for particulate control with a new and modern air quality control (AQC) system for the reduction of sulfur oxides (SOx), particulate matter (PM), and nitrogen oxides (NOx).

The AQC system was commissioned in the 3rd Quarter 2008 and is successfully meeting initial permit emission limits. In this article, the AQC system is described and performance data are given.

It has to be added, that the NOx reduction stage is the first application in the US.

Figure 1. AQC Plant

THE FLOAT GLASS MANUFACTURING PROCESS

The "float process" is the making of flat glass in which the molten glass continuously streams from a furnace tank onto molten tin for forming and is subsequently "annealed" as one continuous sheet. The flat glass is annealed to prevent or remove stresses in the glass by controlled cooling from a suitable temperature. This takes place in an "annealing lehr" which is a long, tunnel-shaped oven for annealing glass by continuous passage. The manufacturing process can be divided into 5 main stages.

1) Raw Material

Flat glass manufacturing begins with the careful weighing and blending of raw materials into a mixed batch. The principal ingredient is silica sand, with others being soda ash, dolomite, limestone, salt cake, rouge, and charcoal. Crushed glass is also mixed into the batch to improve melting characteristics. The receiving, storage, weighing and mixing equipment is enclosed in a fully automated batch house.

2) Furnace

The mixed batch is conveyed to the furnace building where it is continuously fed into the melting furnace at a rate determined in concert with planned production levels. The furnace melts the raw material batch to a state of molten glass with temperatures approaching 1600 °C. The melting furnace is a large refractory structure enclosed in structural and binding steel. Many different types of refractory materials are used in furnace construction. Each one is carefully selected to be used in certain areas where it will perform with a long life and not contribute to product defects.

The melting area of furnace is a region where the glass batch material is melted by energy released from the combustion of natural gas. The combustion process includes the pre-heating of combustion air in structures called regenerators. The regenerators are huge columns of honeycomb-shaped stack of bricks, that is, heat exchangers, through which combustion air flows into the melter 50% of the time and through which the exhaust gasses flow the other 50%. Their purpose on the exhaust cycle is to absorb as much heat as possible from the exhaust gases so that the hot bricks can pre-heat the combustion air on the intake cycle.

Furnace exhaust gases are ducted to the AQC System for SOx, Particulate and NOx emissions reduction before exiting the exhaust stack.

3) Tin bath

Molten glass from the furnace is poured continuously onto a pool of molten tin. The glass floats and spreads across the molten tin reaching a uniform thickness. The molten glass is drawn into a ribbon. Special equipment is used to stretch the glass to the desired thickness. The tin bath is the heart of the float glass manufacturing operation. This is where the molten glass is formed into a continuous ribbon of glass for delivery to the capping line for cutting and packaging.

The molten glass, when flowing on the surface of the molten tin, forms a ribbon of flat parallel surfaces that have the strength and brilliance of sheet glass with optical characteristics and a surface finish superior to plate glass.

4) Annealing Lehr

The ribbon of glass now enters a computer controlled annealing process which precisely regulates the cooling of the glass to room temperature. The cooling process is done in a precisely controlled manner to reduce stresses in the glass which could cause ribbon failure and/or problems in the subsequent cutting operations.

5) Capping Line

After the ribbon of glass leaves the annealing lehr it is trimmed, cut and packaged for shipment to the customer.

AIR QUALITY CONTROL REGULATIONS

Upon completion of the Cold Tank Repair and return to the stable production, the emission control system was required to comply with the flat glass emission limits defined by the San Joaquin Valley Unified Air Pollution Control District in the Table 1.

Table 1: Flat glass emission limits defined by San Joaquin Valley Unified Air Pollution Control District

Category	Limit Relative to Glass Production	US Units	SI Units
NOx emission limits in pounds NOx per ton glass produced	3.25 lb/t rolling 30-day average	242 ppmv	497 mg/m^3 NTP
CO and VOC emission limits – rolling three hour average (ppm limits are referenced at 8% O_2 and dry stack conditions)	300 ppmv CO 100% air-fired furnace 20 ppmv VOC 100% air-fired furnace		
SOx emission limits in pounds SOx per ton glass produced	1.7 lb/t Block 24-hour average (all firing technologies) 1.2 lb/t Rolling 30-day average (all firing technologies)	63 ppmv	183 mg/m^3 NTP
PM-10 emission limits in pounds total PM10 per ton glass produced, block 24-hour average	0.70 lb/t (all firing technologies)	0.04 gr/dscf	107 mg/m^3 NTP
NH$_3$ slip referenced at 8% O_2	0.04 lb/t	10 ppmv	8 mg/m^3 NTP

TECHNOLOGY CONSIDERATIONS FOR EMISSION CONTROL

To meet regulations especially for NOx, Guardian considered emission control options of:

- Oxy-Fuel – DeSOx Scrubber – EP

- DeSOx Scrubber – EP – SCR

Oxy-fuel may be considered a primary method of NOx control at the source, by reducing N_2 introduction to the furnace. Oxygen generation introduced additional considerations of system versus demand, emissions generated off site, and electricity required, as well as overall operating costs of the generation facility. Ultimately, Guardian concluded that oxy-fuel was a costly option to emission control.

Finally Guardian selected SCR DeNOx technology, a technology that is widely used, offers high performance, and is a secondary control technology, controlling thermal formed NOx and batch formed NOx.

AIR QUALITY CONTROL (AQC) PLANT

The gas flow through the AQC plant can be seen from the flow diagram (Fig. 2). The system was designed and supplied by GEA Bischoff.

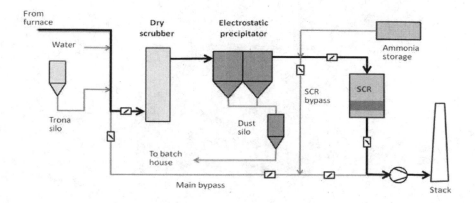

Figure 2: Flow sheet of AQC plant

Furnace exhaust gases in excess of the normal operating temperature range are cooled by water injection prior to the first gas cleaning stage. In the first gas cleaning stage, SOx and condensable particulates are removed. Trona is introduced into the duct prior to the scrubber as a dry sorption reagent. An electrostatic precipitator (EP) removes filterable particulate matter. The captured EP dust is pneumatically transported to an intermediate storage silo for potential reintroduction into the batch. When the EP dust meets quality requirements, it can be recycled, thus eliminating the need to dispose of it as a solid waste.

In the final gas cleaning stage, furnace exhaust gas NOx emissions are reduced in the Selective Catalytic Reduction (SCR) unit. Aqua ammonia is injected upstream of the SCR and NOx is reduced as it reacts with the ammonia on the catalyst bed.

FIRST STAGE: DRY SCRUBBING WITH TRONA

The dry scrubber, a venturi type transport reactor, is designed for three sorption materials: lime hydrate, soda ash, and Trona. Trona® was selected as the primary scrubbing reagent. This mineral is a material of natural origin mined in the US.

Trona® has a unique property providing favorable conditions for the dry sorption of gaseous pollutants.

Exposed to high temperatures, CO_2 and H_2O are liberated from the crystal structure of the Trona® particles. This so-called "popcorn effect" is displayed by the chemical equation and from the SEM pictures of the Trona® before (Fig. 3) and after heating (Fig. 4).

$$2(Na_2CO_3 \cdot NaHCO_3 \cdot 2H_2O) \, (s) \quad \rightarrow \quad 275°F/\, 135°C \quad \rightarrow \quad 3Na2CO3(s) + 5H2O(g) + CO2(g)$$

Figure 3. SEM Picture of Raw Trona (Courtesy of Solvay Chemicals Inc., Houston, TX)

Figure 4. SEM Picture of Heated Trona (Courtesy of Solvay Chemicals Inc., Houston, TX)

As result of the heating at the moment of Trona injection into the hot gas, the particle structure becomes porous, thus giving considerably more surface to the dry sorption of gaseous pollutants such as SO_2, HCl and HF.

Trona® is transported as bulk material by truck to the plant. A storage bin is filled and Trona® is ground and injected into the duct with particle sizes of about 20 μm for reaction with the gaseous pollutants.

Trona injection introduces higher dust load than furnace PM alone, up to 10 to 15 times higher than furnace PM. The Trona further introduces new dust handling problems that can be of a sticky nature. Thus the increased importance of EP design considerations downstream.

SECOND STAGE: PARTICLE COLLECTION

Following the dry scrubber, the Trona® loaded gas enters directly into the EP for particle collection. The EP mechanical design is for high gas temperatures and high dust loads. The EP efficiency is designed for the combined furnace dust plus dry sorbent dust loads.

As the process of electrostatic precipitation is widely known in the glass industry it is refrained to give a deeper description of this dedusting process in this article. However, it should be mentioned that in this specific glass furnace application, there are two EP design details that are worth pointing out due to high and sticky dust loads.

The mechanical form of the collecting plate with zig-zag profile comes close to theoretically ideal current density distribution over the plate, increases the collection efficiency for high dust loads, and its structural rigidity offers higher rapping acceleration for dust removal. This, coupled with the 16" (400mm) spacing of the collecting plates, has been found to be more tolerant of alignment issues of internals and allows for taller collector height with higher input voltage from the power supply.

In addition, an arrangement with two instead of three electrical fields could be selected as a special high performance power supply was used for the first zone; the VARIOVOLT controller system. In comparison to a Thyristor controlled power supply, the controller accommodates the higher dust loads resulting from dry DeSOx Trona injection.

The mechanical design of the EP with zig-zag collectors handles gas temperatures above 800 °F / 430°C, while the overall EP design is for the three sorption materials: lime hydrate, Trona or soda ash.

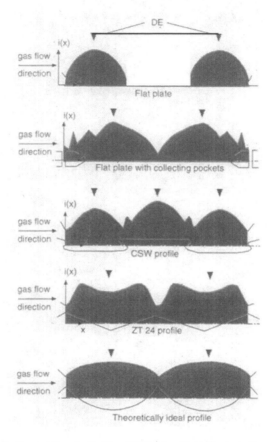

Figure 5. Current Density Profile of Collecting Plate Electrode Geometries

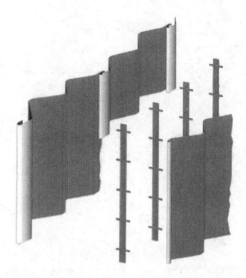

Figure 6: Wide-spacing 16-inch (400 mm) EP Collecting Electrodes

THIRD STAGE: NO_X REDUCTION

The final stage of the AQC plant is the selective catalytic reduction (SCR) of nitrogen oxides which mainly consist of NO (> 95 %) and NO_2. Besides an appropriate catalyst, aqua ammonia is needed as reagent which can be seen from the following reactions.

$$4NO + 4NH_3 + O_2 \rightarrow 4N_2 + 6H_2O$$

$$2NO_2 + 4NH_3 + O_2 \rightarrow 3N_2 + 6H_2O$$

In practice, NO_x is removed by aqua ammonia with the help of a catalyst resulting in nitrogen and water. A uniform distribution of the aqua ammonia in the gas duct upstream of the catalyst is mandatory and is realized by special mixing elements. The arrangement of the SCR unit including ductwork and flow devices can be taken from Fig. 5. A spare volume for another catalyst layer is included in the casing. Flow patterns are optimized by computational fluid dynamics (CFD).

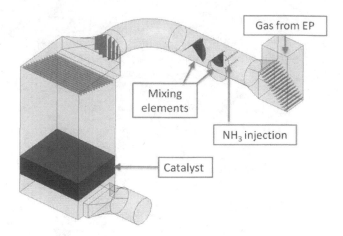

Figure 6. CFD Model For Mixing Optimization

Although the SCR unit is on the clean gas side, it is equipped with an air jet cleaning system to prevent the honeycomb catalyst from long term clogging.

Fig. 7 shows a typical catalyst cartridge prior to installation as well as some dimensions.

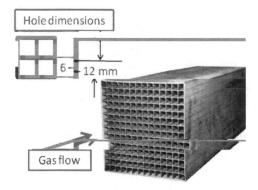

Figure 7. Honeycomb catalyst prior to installation

The structure of the catalyst with its 5mm holes and its clean surface after one year of operation can be seen from Fig. 8.

Fig: 8. Structure of honeycomb – clean surface after 1-1/2 year

EMISSION REDUCTION

After commissioning of the plant in mid 2008, stack emission measurement results show that the emissions are in compliance with the permit requirements. The injected aqua ammonia reagent is reacted with the nitrogen oxides to maintain the ammonia slip below the 10 ppmv compliance limit. It can be mentioned that the EP dust is recycled back into the furnace creating an ideal situation with no dry solid waste or wastewater.

Glass Science, Defects, and Safety

HEAVY METAL ISSUES – IN AND OUT OF GLASS

C. Philip Ross
Creative Opportunities, Inc.
Laguna Niguel, California

The chemistry of industrial glasses is made from multiple oxide components, obtained from a variety of natural minerals which have been beneficiated to consistent physical properties and chemistries. This paper is intended to review certain so-called "Heavy Metals", which are often incorporated into glass; and why glass producers need to understand variables within their manufacturing practices which may relate to a variety of environmental regulations.

This paper in not intended to be an authoritative source for all the current regulations relating to certain heavy metals, or to recommend any specific environmental monitoring or operating practices which may be required to meet both present and future compliance requirements. Rather, this is rather a reminder to industrial glass producers that many metals and metal compounds used in a variety of industrial glass products may be of concern, and that further educational efforts on the issues presented may be required.

There is a tendency to assume that all so-called "heavy metals" and their compounds have highly toxic or eco-toxic properties. To many, this term implies that certain "heavy" metals and all their compounds have the same physicochemical, biological, and toxicological properties, which is untrue.[1]

Heavy metal refers to any metallic chemical element that has a relatively high density (typically at least 5 times the specific gravity of water) and is toxic or poisonous above specific concentrations. OSHA defines Toxic metals, including "heavy metals," as individual metals and metal compounds that negatively affect human health. Examples of heavy metals with specific environmental concerns include selenium (Se), mercury (Hg), cadmium (Cd), arsenic (As), chromium (Cr), and lead (Pb).

The human body has need for approximately 70 friendly trace elements (including some heavy metals) nutritionally essential for a healthy life. Some heavy metals (e.g. copper, selenium, zinc), which in appropriate trace dosages, are essential to maintain the metabolism of the human body. However for some heavy metals, toxic levels can be just above the background concentrations naturally found in nature.

There are another 12 poisonous heavy metals, that act as poisonous interference to the enzyme systems and metabolism of the body. As relatively stable elements (they cannot be metabolized by the body) and bio-accumulative (passed up the food chain to humans) and become a significant health hazard. Heavy metal exposure occurs through three primary routes - inhalation, ingestion and skin absorption. Heavy metals become toxic when they are not metabolized by the body and accumulate in the soft tissues.

---

[1] *Pure Appl. Chem.*, Vol. 74, No. 5, pp. 793– 2002

GLASS HISTORICAL APPLICATION

The following metals (typically in an oxide form) have been intentionally (or incidentally) introduced into a number of glass products for a variety of purposes, and with a range of potential health concerns.

Element	*MWT*	*Purpose*
Arsenic (As)	74.9	Refining Agent
Barium (Ba)	137.9	Glass Former, Sulfate Source
Cadmium (Cd)	113.9	Colorant
Cobalt (Co)	58.9	Colorant
Copper (Cu)	62.9	Colorant
Chromium (Cr)	51.9	Colorant
Mercury (Hg)	200.6	Florescent Light Cullet Contaminant
Manganese (Mn)	54.9	Colorant, Oxidizer
Nickel (Ni)	57.9	Colorant
Lead (Pb)	208.0	Glass Former
Antimony (Sb)	120.9	Refining Agent
Selenium (Se)	79.9	Colorant, Decolorizer
Zinc (Zn)	63.9	Colorant / Opacifier Component

BATCHING ISSUES

Threshold Limit Values (TLV's) are the maximum concentration in air at which it is believed that a particular substance will not produce adverse health effects with repeated daily exposure. It can be a time-weighted average (TLV-TWA), a short-term value (TLV-STEL), or an instantaneous value (TLV-Ceiling). They are typically expressed either as parts per million (ppm) or milligram per cubic meter (mg/m^3).

The batching of glass colorants - such as Se, Co, Ni, Cr, Mn, Cd require- engineering controls, personnel protection, and periodic monitoring to assure compliance with exposure limits. Below are recent examples of some limits which have been set for worker exposure.

Cobalt (Co) - OSHA exposure limit: $0.1 mg/m^3$ for cobalt in workplace air for an 8-hour workday, 40-hour work week. American Conference on General and Industrial Hygiene (ACGIH) occupational exposure limit: $0.02\ mg/m^3$ for cobalt for an 8-hour workday, 40-hour workweek. National Institute for Occupational Safety and Health (NIOSH) occupational exposure limit: $0.05\ mg/m^3$ for cobalt for a 10-hour workday, 40-hour workweek.

Selenium (Se) is the most protective of all nutrients, which is 50 times as potent an antioxidant as vitamin E. It also has the narrowest range between healthful and toxic. Occupational Safety and Health Administration (OSHA) exposure limit in workplace air: 0.2 mg/m3 for an 8-hour day over a 40-hour workweek. The maximum level in toxic waste: 1.0 mg/L.

Chromium (Cr) has three main forms chromium (0), chromium (III), and chromium (VI). Chromium (III) compounds are stable and occur naturally, in the environment. Chromium (0) does not occur naturally and chromium (VI) occurs only rarely. Chromium (III) is an essential nutrient in our diet, but we need only a very small amount.

Nickel (Ni) is considered potential carcinogen. It is typically added as a colorant to glass in the form of an Oxide raw material. The NIOSH recommended exposure limit (REL) on a TWL is 0.015 mg/m^3

Cadmium (Cd) is an extremely toxic metal, and can be used for a glass colorant and certain ceramic decorating paint colors (ACL"s). Standard Number:1910.1027 App A 8-Hour, Time-weighted-average, Permissible Exposure Limit (TWA PEL): Five micrograms of cadmium per cubic meter of air 5 ug/m(3), time-weighted average.
FURNACE / FOREHEARTH VOLATILIZATION

Some raw materials and refractory contain Cr^{+3} can become converted to Cr^{+6} in certain high temperature environments. Chromium (Cr^{+6}) compounds in the workplace air should not be higher than 100 micrograms/m^3 for any period of time. National Institute for Occupational Safety and Health (NIOSH) exposure limit: 500 micrograms/m^3 for chromium (0), chromium (II), and chromium (III) for a 10-hour workday, 40-hour workweek.

In 2006 Occupational Safety and Health Administration (OSHA) amended the standard for occupational exposure to hexavalent chromium. The final rule[2] established an 8-hour time-weighted average (TWA) exposure limit of 5 micrograms of Cr^{+6} per cubic meter of air (5 [mu]g/m^3). This is a considerable reduction from the previous PEL of 1 milligram per 10 cubic meters of air reported as CrO_3, which is equivalent to a limit of 52 [mu]g/m^3 as Cr^{+6} The final rule also contains ancillary provisions for worker protection such as requirements for exposure determination, preferred exposure control methods, respiratory protection, protective clothing and equipment, hygiene areas and practices, medical surveillance, record keeping to meet the new PEL. Identified sources included Chromic Oxide refractories operating at high temperatures, as well as Melters and Forehearths handling Chrome containing raw material feeds.

Lead *(Pb)* - Lead Oxide (like Alkali and Alkaline Earths) is a network modifier in glass. Lead containing glasses include stadium lighting bulbs (~5%), Lead "crystal" tableware (<24%), TV Funnels (~25%), and radiation shielding glasses (40 to 70%). Lead oxide based ceramic glazes have been used for the coating of a variety of ceramic sanitary ware, glass tableware, and home decorative products. Environmental Protection Agency (EPA) limit for lead in air not to exceed 1.5 micrograms/m^3 averaged over 3 months. Occupational exposure to Pb in the glass industry was historically from the furnace exhaust vapors in the manufacturing of CRT glass, which is no longer manufactured in the U.S. If a worker has a blood lead level of 40 micrograms/dL, OSHA requires that worker to be removed from the workroom.

Mercury (Hg) - Soda Lime glass manufactured for Florescent lighting fixtures initially has no mercury, but in use, the inside surface of the glass becomes impregnated with low levels of Hg. If recycled, it must first be properly processed (using a vacuum autoclave technology) to remove Hg before remelting, as its extremely low vapor pressure results in volatilization into the furnace's exhaust.

---

[2]FR February 28, 2006 71:10099-10385

For glass compositions containing regulated heavy metals, there can also be additional concerns with high rates of extraction out of the elevated temperature glass into cullet quench water. This had been a recognized concern for Arsenic containing glasses in the 1960's. The actual amount extracted will vary with type of metal, glass surface area and temperature, and the time the glass is in the water.

FURNACE STACK EMISSIONS

The Clean Air Act Amendment requires the EPA to address the reduction of Urban Air Toxics. Section 112(b) lists 189 Air HAP's, but 33 specific Urban HAP's (the "dirty thirty") are to be specifically addressed in new Area Source Standard regulations for the emitting industries. Facilities which include toxic compounds among raw materials for glass melting are now subject to new standards.

Specific materials subject to regulations for glass include compounds of Arsenic, Cadmium, Chromium, Lead, Manganese, Mercury, and Nickel. Area source standards are to be technology-based and capable of reducing emissions by ~ 90 %. In 2007 EPA adopted the requirement for melting furnaces to meet the Federal New Source Performance Standard (NSPS)[3] for particulate (typically 0.2 lb. per ton of glass melted), or perform stack tests using EPA's M-29 (Determination of metals from Stationary Sources) tests to confirm emissions are below 0.02 lb. total HAP's per ton of glass melted.[4]

Because of low vapor pressure, many heavy metals are highly volatile at glass melting temperatures and a significant percentage of their batch input does not stay in the glass being melted. Consequently, the vapors may find specific condensation temperatures within the exhaust system. These can build up to significant concentration in refractory lining and flue debris over the life of furnace campaigns. Consequently, concerns with worker exposure during flue cleaning, furnace demolition and the disposal characterization of such debris must be properly recognized.

PRODUCT USE IN SERVICE

Parts per million (ppm) measurements of heavy metals can be found in most container glasses. Most of these heavy metals are not the part of any batch formulation with intentional additions, but rather the result of incidental presence in the raw materials. Other than trivalent Chromic Oxide (Cr_2O_3) which is an intentional colorant for green glasses, the source of the other heavy metals are most typically from post consumer cullet being recycled.

The presence or heavy metals in container glass does not automatically mean there are health concerns. Commercial glass containers meet the USP Type III requirements for chemical durability, which means very little of the surface of the glass will interact with the containers content. The overall glass chemistry, the nature of the product in the container, time and temperature will determine the extent of any heavy metals leaching out of the glass structure and into a food product. Specific tests are available to obtain objective measurements. Some glass users have intentionally limited the level of heavy metals to less than 100 ppm, and also specify specific leaching tests to determine the extent any of these heavy metals can come out of the glass.

Glass produced with no recycling content will have very low levels of heavy metals, while high recycling content (particularly with the above mentioned sources) will have much higher levels of

---

[3] 40 CFR 60, subpart CC
[4] 72 FR 73180 Dec 26 2007

heavy metals. Some imported glass containers have higher levels of lead than those produced in the US. Lead free ACL paints have become more common in the past decade, and over time will result in lower Lead and Cadmium levels in recycled glass.

Model Toxics in Packaging Legislation (MTPL) was developed to reduce the amount mercury, lead, cadmium, and hexavalent chromium used in all packaging. The intended purpose was to prevent the use of toxic heavy metals in packaging materials that enter landfills, waste incinerators, recycling streams, and ultimately, the environment.

The Toxics in Packaging Clearinghouse (TPCH) was formed in 1992 to promote the Model Toxics in Packaging Legislation. This model legislation was originally drafted by the Source Reduction Council of the Collation of North East Governors (CONEA) in 1989. The Clearinghouse includes a states-only voting membership and an industry/public interest advisory group. The legislation has been successfully adopted by nineteen states: California, Connecticut, Florida, Georgia, Illinois, Iowa, Maryland, Maine, Minnesota, Missouri, New Hampshire, New Jersey, New York, Pennsylvania, Rhode Island, Vermont, Virginia, Washington and Wisconsin. Other states, as well as the U.S. Congress, have also considered the legislation.

According to the legislation, companies are not permitted to sell or distribute any package or packaging component to which any of the four metals has been intentionally introduced. Packaging components may include coatings, inks and labels (such as ACL's). Companies are given two years after the enactment of the legislation to clear inventories and reformulate their packaging. The law further requires that the incidental presence of the metals be gradually reduced to 100 parts per million four years after it has been enacted. Incidental presence is defined as the presence of one of the four regulated metals as an unintended or undesired component of the final package.

Since laboratory testing was relatively expensive, states' ability to broadly test market samples of packaging went largely unnoticed until 2006, when the TPCH initiated its first screening project (using x-ray fluorescent analyses for the presence of heavy metals). In 2007, TPCH released the first comprehensive report on heavy metals in packaging, based on XRF screening. Heavy metals restricted by state Toxics in Packaging laws, particularly lead and cadmium, were frequently found in some types of packaging and packaging components, particularly imports.

In 2008, TPCH screened an additional 409 packages to detect trends in compliance with Toxics in Packaging laws and identify areas where TPCH should focus, or continue to focus, its outreach efforts. TPCH used the screening results to notify brand owners of potentially non-compliant packages about toxics in packaging requirements, and to bring non-compliant packages into compliance. The 2009 update[5] documented the continued investigation by TPCH of heavy metals in packaging, using XRF analysis.

APPLIED CERAMIC LABELS (ACL)

Ceramic and glassware are often decorated with permanent, kiln-fired colors and glazes that have traditionally contained heavy metals. The use of these metals, primarily lead and cadmium, has come under significant examination relative to environmental and human health concerns. When in contact with foods and beverages, the metals can leach out of the decorations. Prolonged or repeated food contact could result in increased health risks. Ceramic or glass packaging that enters the waste stream

---

[5]The Toxics in Packaging Clearinghouse - An Assessment of Heavy Metals in Packaging: 2009 Update

can also leach components over time, which can have consequences on the environment and water supplies.

In response to these concerns, there are a number of regulations limiting the amount of heavy metals in food-bearing vessels and disposable packaging. The Federal Food & Drug Administration (FDA) has Compliance Guides setting forth standards for the leaching of lead (CPG 7117.07) and cadmium (CPG 7117.06) from the food contact surface of ceramic and glass tableware. A voluntary industry standard limits the amount of lead and cadmium leaching from the top 20 mm of the outside of a ceramic cup, mug, or drinking glass. The limits are not more than 4 ppm of lead and not more than 0.4 ppm for cadmium.

There are a variety of individual state laws and voluntary industry standards. As example, *California's Proposition 65 or* "Safe Drinking Water and Toxic Enforcement Act" requires that tableware leaching lead in excess of the standard set forth by the regulation have proper warnings when they are sold or displayed. The manufacturer must inform retailers when warnings are required and provide the appropriate signs. Warning notices are required of lead leaching in excess of 0.226 parts per million for flatware and 0.100 parts per million for hollowware, cups, mugs, and pitchers. All products that may contain harmful substances under these definitions are required to have a warning placed on them.

When glass containers with ACL are recycled, the ceramic oxide materials and colorants from the label become incorporated into the new glass being produced. Lead Oxide had been in Ceramic decoration labels ("ACL") to provide a lower softening point glass to be fused onto the surface of normal glass containers. Cadmium was used to provide Yellow, Orange, and Red in these labels. Lead has also entered the recycling stream from metallic foils from wine bottles, lead end caps from florescent light bulbs, and lead crystal assumed to be compatible with recycling.

ASTM (American Society for Testing and Materials International) established the C1606-04 Standard Test Method for Sampling Protocol for TCLP (Toxicity Characteristic Leaching Procedure) *Testing of Container Glassware*. This procedure outlines the proper preparation for leaching tests on container glassware for accurate results. The maximum allowed amounts of individual heavy metals as a result of the TCLP test are: 1.0 ppm (parts per million) lead, 5.0 ppm cadmium, and 5.0 ppm hexavalent chromium.

Properly vitrified glass or ceramic decals are no longer considered to be a separate packaging component. The finished glass or ceramic package must pass the leaching test as described in the U.S. EPA Document "Test Methods for Evaluating Solid Waste, Physical/Chemical Methods", also known as SW-846. Most recently, EPA has taken the position that no intentional addition of Cadmium will be allowed, in spite of such labels passing other leach standards.

The Consumer Product Safety Improvement Act of 2008. stipulates that after August 14, 2009, manufactured products for children cannot contain more than 300 ppm of lead. After August 14, 2011, if the methods exist to make it feasible, the most amount of lead able to be contained in these products will be 100 ppm. Under this law, manufacturers, retailers, and importers can all be held liable.

FACILITY DISPOSAL / LANDFILL

Landfill regulations typically have concerns with levels of heavy metals leaching out of glass or other material (such as furnace rebuild refractory debris). States and regional authorities define specific content differences between "hazardous" and "non-hazardous" waste, as well as the appropriate test procedures for characterization. "EPA's Toxicity Characteristic Leaching Procedure[6] (TCLP) is designed to determine the mobility of both organic and inorganic chemicals present in liquid, solid, and multiphase wastes. If an analysis of the TCLP extract indicates that any regulated compound is present at such high concentrations are above the regulatory level for that compound, then the waste is hazardous. Specific concerns have typically focused upon Chrome containing refractories. But often flue condensate can have suprisingly high levels of heavy metals deposited over years or decades.

CONCLUSIONS

There are a variety of heavy metals which either intentionally or incidentally may become part of glass products or the manufacturing environment. Glass manufacturers must identify the presence of all heavy metals in their operation and become familiar with ever increasing levels of regulatory standards, which relate to a variety of manufacturing practices in the glass industry. These includes employee exposure to batching and process emissions within their facility, as well as emissions from melting operations. All of which will require engineering and procedural controls. Additional concerns may relate to use in service and ultimate disposal of their glass products.

---

[6]EPA Method 1311

A LOOK AT THE CHEMICAL STRENGTHENING PROCESS: ALKALI ALUMINOSILICATE
GLASSES VS. SODA-LIME GLASS

Sinue Gomez, Matthew J. Dejneka, Adam J. Ellison, and Katherine R. Rossington
Glass Research, Corning Incorporated, Corning, NY, USA

ABSTRACT

Alkali alumino silicate and Soda-lime type glasses can be strengthened via ion-exchange. Compared to soda-lime glasses, alkali aluminosilicates can achieve higher compressive stresses and deeper compressive layers which results in superior strength. High damage resistance and strength are required in glass for applications such as consumer electronics which are exposed to potentially damaging stresses and impacts in everyday use. The effect of time, temperature, and ion-exchange bath chemistry on the chemical strengthening of alkali aluminosilicate glasses and soda-lime type glasses will be discussed.

1. INTRODUCTION
The Automobile windshield industry was the initial incentive for the commercial application of chemically strengthened glass. The ion-exchange process has been widely used in the fabrication of missile nose cones and cladding for aircraft windshields. Chemical tempering has also been used in high end photochromic and white crown eyeglass lenses, and in the production of hard disk drives for computer data storage.[1,2]

Relative to plastic, glass is not only much harder and tolerant to damage, but also offers superior optical quality and a richer look which is highly desirable in consumer electronic applications. Thus, the possibility of strengthening glass sheets allows using glass in applications, such as cover plates for cellular phones, PDAs, notebooks, etc., where high strength and scratch resistance are necessary due to everyday exposure.

Theoretically glass can be as strong as ~14 GPa.[3] However, the presence of flaws or defects in glass can drastically affect its strength. In practice, strengthening processes in glass mainly prevent surface flaws or cracks from propagating when external forces are applied to the surface. Since glass is stronger in compression than in tension, the introduction of surface compressive stress profiles in glass is a well known approach for strengthening. The ion-exchange process or chemical tempering of glass can be used to incorporate residual stresses by exposing a glass containing alkalis to molten salt baths that contain alkali ions which are typically larger than those initially in the glass. As a result of chemical potential differences, some ions in the glass are replaced by ions in the molten salt bath. If a smaller ion in the glass is replaced by a larger ion from the bath, a compressive layer is formed in the glass surface producing compressive stress (CS). This compressive stress is balanced by a volume below the surface under tensile stress (CT).

The effectiveness of the compressive layer depends on the size of the initial flaws, the depth of compression, and the magnitude of the compression at the surface. All silicate glasses can be chemically tempered provided that enough mobile cations are available in the composition. However, glass composition plays a very important role on the success of chemical tempering. For a given glass composition, time, temperature and salt bath composition are crucial during ion exchange (IX) for obtaining desired properties. The effect of time, temperature, salt bath chemistry on IX ability of different glass compositions and resulting mechanical properties will be discussed.

2. EXPERIMENTAL

Alkali aluminosilicate glasses, Gorilla® glass (GG), and soda-lime type glasses (SLS) were investigated. Samples of dimensions 50mm x 50 mm x 1mm were exposed to molten KNO_3 or $KNO_3/NaNO_3$ mixtures at temperatures between 380°C and 430°C for times ranging from 2h to 16h. The depth of layer (DOL) and CS were determined using a surface stress meter FSM 6000. In some instances DOL was determined optically via polarimetry.

Potassium diffusion profiles were determined by electron microprobe analysis (EPMA). EPMA line scans were performed on a JEOL 8900 Superprobe using a 15 keV, 20 nA focused beam.

Mechanical tests were performed on ion-exchanged SLS and GG. Damage tolerance or retained strength was measured via ring-on-ring (ROR).[4] Damage was induced via scratching the glass surface with a Vickers indenter at a controlled load and rate.

3. RESULTS AND DISCUSSION

3.1 Glass Composition

Glass composition plays an important role on determining IX properties. Different glass compositions require different IX times and temperatures in order to obtain a given CS and DOL. Time and temperature are critical for obtaining appropriate stress levels in a particular glass. Figure 1 shows the relationship between DOL and CS for a set of SLS-type glasses compared to GG and a series of alkali aluminosilicates. As time and temperature increase, DOL increases, while CS decreases. In the range of times and temperatures explored, GG along with other alkali aluminosilicate glasses show larger DOL and CS compared to SLS glasses. To achieve a DOL of ~30 microns, only 6 hours is necessary for GG; while in most commercial SLS glasses the time required is at least 16h. Against SLS, GG is not only capable of achieving deeper DOL in shorter times, but also maintains higher CS. The longer time and higher temperature required to achieve deeper depths of compression in SLS glasses are detrimental for maintaining high CS. Notice that the change in CS as DOL increases is significantly larger for SLS glasses (~200MPa) compared to GG (~100MPa). This larger change in CS is a result of the higher rate of relaxation in SLS glass.

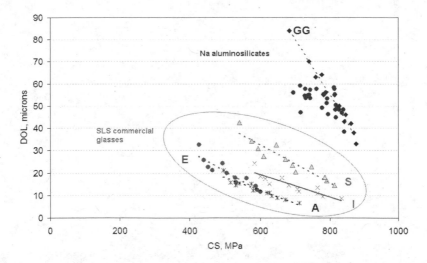

Figure 1. DOL vs. CS for a set of commercial SLS-type glasses, GG, and other Na aluminosilicate glasses ion-exchanged at 390-430°C for 6-16h.

Large depths of compression can be obtained in alkali aluminosilicate glasses containing lithium. The rate of diffusion when replacing Na^+ for Li^+ can be ~10x faster than substituting K^+ for Na^+. Table 1 shows the DOL and CS obtained for aluminosilicate glasses having between 3-9 mol% lithium that were ion-exchanged at 380°C in a mixed salt bath containing 60wt% KNO_3 and 40wt% $NaNO_3$. Smaller ions diffuse faster in some glasses than larger ions, thus larger depths of compression can be achieved; however, the achievable CS in lithium aluminosilicates (Figure 2) is considerably lower compared to that of ion-exchanged sodium aluminosilicates (Figure 1). Larger ions, on the other hand, can produce higher compressive stresses than smaller ions. Large depth of layer is desirable since it imparts protection of the glass surface from flaws or damage, while CS imparts strength to the part. Thus, determining the desired properties on a particular application can be crucial to choose a suitable ion-exchangeable composition as well as process conditions.

Table 1. DOL for lithium containing alkali aluminosilicate glasses ion-exchanged at 380°C in a 60wt% KNO_3 and 40wt% $NaNO_3$ molten salt bath.

Glass	Time, h	DOL, µm
Glass E	4	50
Glass E	8	106
Glass F	4	50
Glass F	8	100

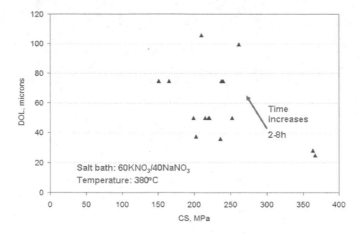

Figure 2. DOL vs. CS for several lithium aluminosilicate glasses ion exchanged in 60KNO₃/40NaNO₃ for 2-8h at 380°C

3.2 Salt Bath Chemistry

Salt bath chemistry is critical for obtaining desired IX properties. The amount of compressive stress which can be realized in an ion-exchangeable glass will depend on the concentration of ions present in the salt bath. In sodium aluminosilicate glasses, additions of $NaNO_3$ to a KNO_3 salt bath will affect the concentration of K ions at the bath-glass interface, reducing the amount of K^+ available for strengthening and impacting the levels of stress in the glass. Figure 3 shows the relationship of the concentration of K at the glass surface relative to the concentration of KNO_3 in the salt bath.

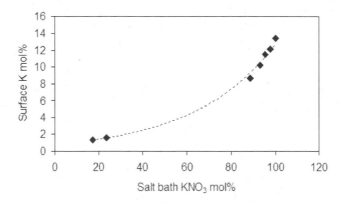

Figure 3. Relationship between salt bath chemistry and surface chemistry in IX glass at 410°C for 7.5h

Since the K_2O concentration at the surface (C_o) decreases as $NaNO_3$ increases in the salt bath, consequently the compressive stress at the surface also decreases. The impact of surface concentration for samples IX for 9h at 410°C in salt baths of different compositions is shown in Figure 4, where the addition of 14wt% sodium results in a CS decrease of ~40% percent.

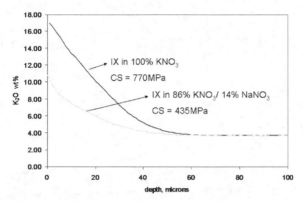

Figure 4. K_2O diffusion profiles of an alkali aluminosilicate glass IX at 410°C for 9h utilizing different salt bath compositions

The presence of sodium in a potassium salt bath has a strong impact on the levels of stress of IX glass, while DOL is not greatly affected. This indicates that the diffusion rate is not limited by Na in the molten salt bath. Similarly, in Li-containing alkali aluminosilicates where the Na for Li exchange is desirable, the presence of Li in the salt bath, typically $NaNO_3$ or mixtures of $NaNO_3$ and KNO_3, will be detrimental for obtaining the needed compressive stress and even worse, the reverse Li for Na exchange is known to occur and such an ion exchange tends to produce a tensile stress in the glass surface which will weaken the glass. Since, Na for Li exchange occurs at a much higher rate than K for Na exchange the speed at which the salt bath gets "contaminated" is faster.

3.3 Mechanical properties
Large DOL and high CS are desirable since it imparts protection of the glass surface from flaws or damage. If damage resistance is critical, a glass able to produce large DOL is recommended. On the other hand, if unhandled or pristine strength is the main driver of the application, then CS can become more important. Ultimately, a combination of both large DOL and CS could be ideal. Temperature and time allow controlling the rate of ion diffusion and the extent of such diffusion, and as a consequence the magnitude of the stresses and DOL.
Figure 5 compares the retained strength via ring on ring (ROR) of IX GG and SLS after scratching the surface at various loads. The results show a steep drop in strength in SLS glass with loads as little as 0.25 N compared to GG which retains its strength after 1 N scratch loads. The slower drop in strength for GG glass denotes its greater damage tolerance when compared to SLS. The depth of the compression layer or DOL plays an important role in determining damage tolerance. The DOL of commercially available SLS glasses ranges between 9-15 microns; while GG has a DOL larger than 40 microns. The deeper DOL enables GG to retain its strength after scratching with a Vickers indenter and 1 N load, unlike SLS which has lost its strength because the induced flaws penetrated its compressive layer.

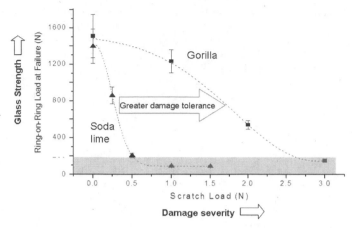

Figure 5. GG vs. SLS ROR load to failure as a function of damage induced by scratch.

4. CONCLUSIONS

Ion-exchange of glass opens a great deal of opportunities to implement glass in applications where damage resistance and toughness are priorities, such as cover glass for consumer electronic applications. Glass composition as well as the ion-exchange process parameters can be manipulated to obtain mechanical properties to suit given applications. For a given glass composition, time, temperature and salt bath composition are critical factors in obtaining appropriate depths of compression and stress levels for a particular glass and application.

Compared to commercially available SLS-type glasses, alkali aluminosilicate glasses can be ion-exchanged to larger depths of compression and higher surface compressive stresses in shorter amounts of time. The deeper DOL and higher CS achievable by GG impart superior strength and damage tolerance. This fact is of great importance in the consumer electronics applications where the glass is used as a cover and everyday use exposes the glass surface to damage.

5. ACKNOWLEDGMENTS

We would like to thank Dr. Ben Hanson for microprobe analysis. We also want to thank Dr. Ron Stewart for providing mechanical data.

6. REFERENCES

[1] Kellman, C., Glass Industry, 1993, 74, 23-24
[2] Gy, R., Mat. Sci. and Eng. B, 2008, 149, 159-165
[3] Glass Science and Technology, Volume 5, Edited by Uhlmann, D.R. and Kreidl, N.J. Academic Press, 1980, New York, 133-26
[4] ASTM-C1499 2009, Standard Test Method for Monotonic Equibiaxial Flexural Strength of Advanced Ceramics at Ambient Temperature

STUDYING BUBBLE GLASS DEFECTS THAT ARE CAUSED BY REFRACTORY MATERIALS

Jiri Ullrich and Erik Muijsenberg
Glass Service Inc., Vsetin, the Czech Republic

Glass Service, Inc. (GS) hereby presents a paper entitled, "Studying Bubble Glass Defects that are caused by Refractory Materials." Various laboratory measurements will be discussed to determine the best approach for identification of the gas bubble defects in the glass, with a focus upon those defects potentially arising from refractory sources.

First, though, it is important to note that there are many likely sources of gas bubble defects in the glass, arising from one or more of the following sources:

1) Improper thermal or melting curve within the furnace
2) Insufficient fining or achievement of sufficient melting temperatures for a sufficient period of time
3) Electrical grounding issues, typically caused by a faulty thermocouple
4) Glass re-boil from reheating of the glass
5) Contamination of the glass melt by various materials, whether refractory or other
6) Mechanical entrapment of air or furnace atmosphere, typically by stirrers
7) And, bubbles caused by refractory.

Among the aforementioned causes of defects, GS has determined that a majority of the bubbles that are analyzed are related to refractory issues.

Root causes for the determination of refractory-related gas bubble defects arise from the analyses of many different glass production bubbles, and observation made through our High Temperature Observation (HTO) furnace. Many of the gas bubble defects analyzed indicate bubbles of a similar nature to that of a refractory-caused defect. Furthermore, a direct analysis of the gaseous composition of the bubble defects will be reviewed, but again resulting in refractory-related defects.

Gas bubble defects can be analyzed utilizing supplemental laboratory analysis by the HTO method:

a. The first video illustrates the bubbles released from the refractory surface. The quantity and rate of gas bubbles that are released from the refractory can be determined and subsequently analyzed. (Analyses of the bubbles are done utilizing a mass spectrometer to determine the composition of the types of internal gases, the internal pressure, and any deposits.)

b. A comparative method can also be utilized whereby two (2) individual pieces of refractory of different types can be placed in the same HTO crucible melt. The quantity and release of gas bubble defects can be comparatively analyzed. (In this case, it should be noted that bubbles are not analyzed by the mass spectrometer due to potential co-mingling of this gas bubble defects.)

c. A semi-quantitative evaluation at specified conditions is also suggested, including different temperatures.

An illustration of the HTO melting process is illustrated in Figure 1 below.

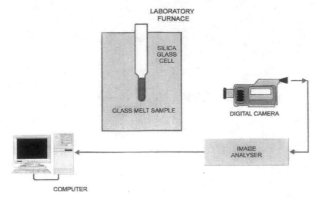

Figure 1

Basically, the glass sample including the refractory is placed in a quartz (silica) tube. This tube is then inserted into an electrically heated furnace. (Note that the atmosphere over the glass melt can be simulated to reproduce the same melting conditions as found in either an oxy/fuel furnace or an air/fuel furnace.) The silica tube is then preheated to a pre-defined temperature, usually 1200°C up to 1600°C, whereby a video image of the melting process is recorded. The digital camera records time lapse images of the melting process. Subsequently, via an image analyzer, one can determine the number of bubbles present in each video sequence, such that a graph of the number of bubbles over time can be plotted.

The number of bubbles released over time is an important indicator of the start-up time for gas bubble defect generation. Furthermore, the total refractory area beneath the glass melt level can be compared for quantitative analyses of the number of gas bubble defects that are generated from different samples. Finally, the end result or background amount of bubbles that remain present from the refractory source can be determined and compared between different refractories. Essentially, one can get a quantitative comparison of different glass melts with different refractories and determine the time, rate, quantity, and background level of gas bubble defects.

Shown in Figure 2 (below) is an HTO furnace with an open air atmosphere above the glass melt. Note the insertion and suspension of the silica tube from the top of the HTO furnace and the side or horizontal view of the camera that records the video images of the glass melt during the melting process over time.

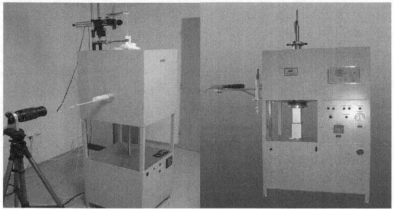

Figure 2   Figure 3

Figure 3 (above) illustrates the HTO furnace with the vacuum option, or the individual control of the furnace atmosphere above the glass melt. In other words, the furnace can simulate an oxy/fuel furnace firing condition, a typical air/fuel firing condition, or either an oxidizing or reducing condition that may be present above the glass melt.

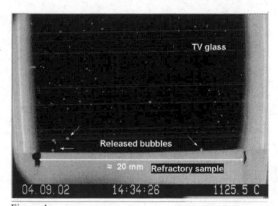

Figure 4

Figure 4 (above) illustrates the release of gas bubbles that creates defects from a refractory sample in the bottom of the silica crucible. The glass melt in this case is for TV panel glass, and the melting temperature is 1125°C, which is indicative of refractory in the working end / forehearth area of the furnace.

A video image shows an initially higher concentration of gas bubbles released into the glass melt, followed by a gradually declining number of bubbles. The principal aim of the test was to determine the number of bubbles released for two (2) different types of refractories at the same melting temperature with the same glass melt composition.

Figure 5 (below) illustrates a melting temperature of 1350°C, and shows a melter bottom refractory that approximates the melting temperatures within the furnace. The illustration denotes the gas bubble formation.

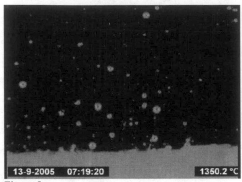

Figure 5

The process is mostly semi-quantitative, since many of the bubbles released from the refractory / glass interface are evaluated by an image analyzer comparison between various test melts. For example, the visual comparison of the melts is recorded, whereby a bubble count on the surface / glass interface can be made. In the case of higher quantities of the released bubbles, bubble counting within a selected area above the refractory surface / glass interface can be made. Finally, the bubble / glass melt ratio above the glass surface can be determined. Essentially, the comparison is between the bubbles (white dots) contrasted against the melted glass (black background).

Figure 6 (below) shows a plot of the quantitative number of bubbles that are produced from the refractory surface over time versus an isothermal course at 1160°C. From the figure, the best material is the first Fused Cast Alumina followed closely by the Fused Cast AZS, then the Sillimanite, then finally the second Fused Cast Alumina.

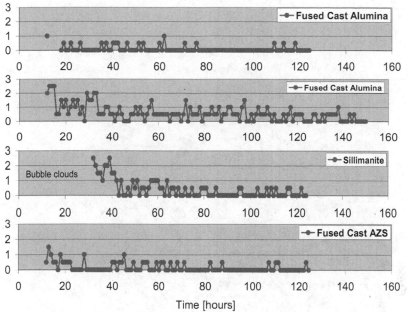

Figure 6 – Given as N [quantity.cm-^{2}hour-^{1}]

Figure 7 shows four (4) different refractories: Fused Cast Alumina; a first Sintered Zircon-Mullite sample (red line) containing a higher amount of zirconium oxide per unit then; a second Sintered Zircon-Mullite sample; and a Bonded Alumina. Shown is the number of bubbles above the refractory surface as it is exposed at a temperature of 1200°C in hours. By far, the second Sintered Zircon-Mullite material has the greatest number of bubble defects.

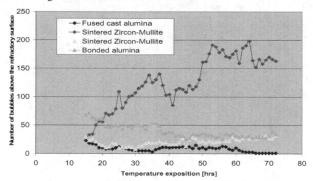

Figure 7

Figures 8 and 9 (below) show the effect of an increased temperature on the evolution of bubbles for the four (4) aforementioned materials. The HTO test was started at 1200°C and increased to 1340°C resulting in a significantly increased number of bubbles.

Figure 8 shows the melting images of the HTO refractory test on bubble evolution after two hours with a temperature increase to 1340°C. The most drastic release of bubbles occurred with the Bonded Alumina. That being said, the first Sintered Zircon-Mullite also had a large increase. The second Sintered Zircon-Mullite and the Fused Cast Alumina had the lowest increases.

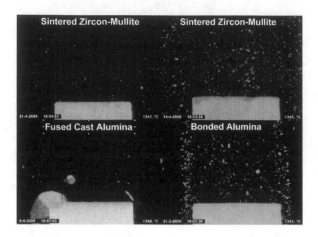

Figure 8

While the previous Figure 7 represented only the beginning phase of the tests, the following Figure 9 shows the entire course of the test, when a temperature increase to 1340°C was subsequently applied. It is interesting that this first Sintered Zircon-Mullite (red), the poorest performing refractory sample at 1200°C, had a significantly reduced bubble release at 1340°C.

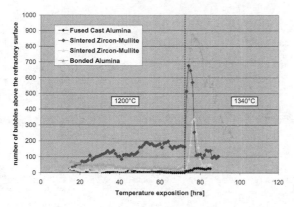

Figure 9

The use of the HTO test furnace to provide a visual image of the refractory gas bubble evolution is good for the beginning phase of the test. The test can last from three (3) to five (5) days depending upon the testing temperature, and it directly correlates to corrosion of the silica crucible and subsequent glass leakage or crystallization of the silica crucible wall. Basically, the glass sidewall transcends from a transparent to an opaque viewing surface.

If an extended time period is required, the refractory sample can be tested in a platinum crucible with a time duration of up to several weeks, or longer. Of course, a side visual observation can not be made, but the sample can still be cooled down and analyzed with the mass spectrometer to determine the gas bubble composition in the glass melt.

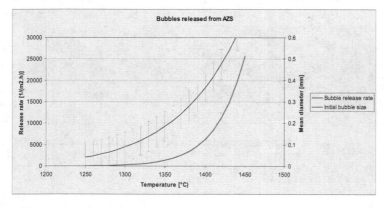

Figure 10

An HTO test was conducted as shown in Figure 10 (above) to indicate the release rate and the initial size of the bubbles as a function of temperature. In the example, bubbles are released from the AZS refractory with the mean diameter plotted on the right-hand "y" axis while the release rate is shown on the left-hand "y" axis. The initial bubble size and bubble release rate is shown as a function of temperature on the "x" axis. Initial size and bubble release rate are an exponential function of increased temperature.

Figure 11 (below) illustrates an animation by mathematical modeling of the formation of gas bubble defects within the furnace. Included is a float furnace example, including the melter, refiner, waist, working end, and up to the canal.

It is generally assumed that the refractory bubble simulation is uniform for any given refractory, however, the rate of bubble release from the refractories increases with increased temperatures. Simply as a function of the furnace operation, refractories near the output are usually much colder than the refractories upstream. The melting area of the furnace is generally 1550°C whereby the colder areas of the furnace are as low as 1250°C to 1300°C. Because of the melting process, there is a major release of bubbles upstream. Hence, a uniform initial distribution of the gas bubble defects is not realistic, (i.e. some refractory locations are relevant, while others produce many bubbles.)

Preliminary results – New method

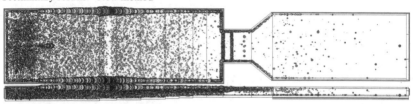

Older method

Figure 11

Utilizing the data from the initial bubble size and bubble release rate, the bubbles as released from the AZS material can be included within the mathematical simulation of the furnace. A bubble trace program is incorporated within the mathematical modeling software, and by determining a particular furnace operation and its operating temperatures, the release of refractory bubbles can be simulated in the model.

Previously, bubbles were introduced into mathematical models in a uniform way, which yielded imprecise results. However, today's results of the initial bubble size and the bubble release rate as a function of temperature can be much more accurately modeled.

Also shown in this figure is the furnace operation with the bubbles released from the refractory, which illustrates a significant amount along the glass contact sidewall blocks, with a reduction over the furnace operation and into the working end. Most bubbles are refined in this particular operation.

Another comparison is made in Figure 12 (below) by utilizing the bubble tracing procedure within the model, where the bubbles reaching areas which potentially cause defects in the furnace output canal are tracked using both the traditional method and also by the new mathematical model. The traditional method shows evenly distributed and random sizes of bubbles, while the new mathematical model is based on experimental results, where the release rate and size as a function of temperature are considered.

From the same figure, it can be seen that there is not much difference on release rate of the bubbles potentially causing defects adjacent to the glass contact refractories near the batch melting area, but when contrasted between the refiner and the waist, there is a significant differential. This differential carries well into the working end as well. The initial bubble size and the bubble release rates show a much higher fining capacity due to experimental results and their incorporation into the upgraded mathematical model studies.

Traditional Method – evenly distributed, random size

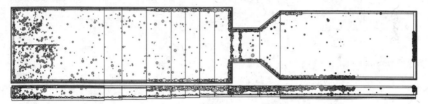

Based on the experiment – release rate and size as functions of T

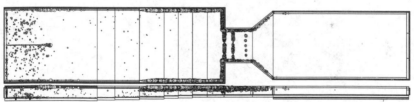

Figure 12

Next, we wish to illustrate the analysis of the gas bubbles generating glass defects. To do so, gas bubbles need to be melted in a separate crucible, since the bubbles generated from within the silica crucible from the HTO experiment are generally destroyed after cooling down. In other words, a separate crucible melt can be undertaken with the cool-down procedure and bubble analysis made subsequently.

With the crucible melt, the true refractory bubbles can be analyzed since there is molten glass and refractory only. Providing there is no overheating or reboil, the bubbles that are generated are, in fact, those generated by the refractory.

Figures 13a, 13b, and 13c (below) represent different views of the components of a mass spectrometer. Shown in Figure 13a is the analytical processor for the analysis of the glass bubble defects.

Figure 13a

Figure 13b shows the mechanical breaking chamber for the glass defects, in addition to the evacuation chamber and analytical devices.

Figure 13b

Figure 13c shows the sample preparation for the gas bubble defect analysis. It can be seen from the overall dimensions of the sample that it is quite small to fit within the breaking and analysis chamber.

The dimensions are generally 5 mm wide, 15 mm long, and 3 mm in depth. (The maximum dimensions are 10 mm wide, 55 mm long, and 10 mm in depth.)

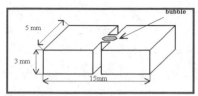

Figure 13c

Shown in Figure 13c, the gas bubble defect is located within the center of the defect with notches or grooves inscribed to either side of the glass. Later, when pressure is applied to the piece of glass, the breakage of the glass is through the center of the gas bubble defect. The gases are then released whereby the mass spectrometer determines the overall pressure of the gases within the bubble, as well as the gas composition.

Figure 14 (below) gives the gas analyses of bubbles released from various types of different refractories.

Sample ID	Dimension [mm]			D.EQ. [mm]	Volume [ml]	p [kPa]	Gas composition [%]					
	a	b	c				N2	CO2	O2	Ar	SO2	H2S
Fused cast AZS (40% ZrO$_2$)	0,32	0,32	0,30	0,31	1,57E-05	30,5	81,6	17,5		0,9		
	0,20	0,20	0,20	0,20	4,19E-06	2,5	81,1	18,0		0,9		
	0,40	0,40	0,38	0,39	3,18E-05	30,0	74,3	25,1		0,6		
	0,40	0,40	0,40	0,40	3,35E-05	32,0	76,2	21,5	1,6	0,7		
Fused cast AZS (40% ZrO$_2$)	0,17	0,17	0,17	0,17	2,40E-06	35,5	55,8	43,8		0,4		
	0,22	0,22	0,22	0,22	5,28E-06	35,5	37,4	62,4		0,2		
	0,50	0,50	0,50	0,50	6,54E-05	30,0	83,4	16,0		0,6		
	0,53	0,53	0,53	0,53	7,80E-05	34,0	53,4	46,1		0,5		
Fused cast alumina	0,33	0,33	0,33	0,33	1,88E-05	7,0	64,5	34,0	0,8	0,7		
Fused cast AZS (41% ZrO2)	0,13	0,13	0,13	0,13	1,06E-06	37,0	62,1	37,1		0,8		
	0,20	0,20	0,22	0,21	4,52E-06	38,0	73,1	26,2		0,7		
Fused cast zirconia	0,17	0,17	0,17	0,17	2,40E-06	39,0	14,6	85,4				
	0,17	0,17	0,17	0,17	2,40E-06	40,0	14,2	85,8				
	0,30	0,28	0,27	0,28	1,17E-05	50,0	9,9	90,1				
Fused cast AZS (36,5% ZrO2)	0,42	0,42	0,21	0,33	1,93E-05	34,0	15,2	84,7		0,1		
	0,40	0,38	0,38	0,39	3,02E-05	40,5	13,5	86,4		0,1		
	0,33	0,33	0,33	0,33	1,88E-05	41,5	9,5	90,5		TR		
Sintered alumina (93% Al$_2$O$_3$)	0,12	0,12	0,12	0,12	9,05E-07	34,0	73,7	26,0		0,3		
	0,15	0,15	0,15	0,15	1,77E-06	27,5	64,0	35,5		0,5		

Blue - Air residuals   Yellow ~Carbon containing impurities

Figure 14

Most of the gas bubble defect sizes range in a diameter equivalent of approximately 0.2 mm up to 0.4 mm. Additionally, the internal pressure of the gas bubbles was generally higher at 30 to 40 kPa (kilopascals). That being said there were some differences between the nitrogen and carbon-dioxide levels between the various refractories, but in most cases there were some corresponding level of Argon which indicates a penetration of air from within the refractories into the glass melt.

Most important, the Fused Cast Zirconia and the Fused Cast AZS (36.5% zirconia oxide) had some carbon containing impurities because the level of nitrogen was quite low and the level of CO_2 quite high. It was generally between 10% and 15% nitrogen, and 85% to 90% CO_2.

On the other hand, the Fused Cast Alumina, Fused Cast AZS (40% and 41% zirconia oxide), and the Sintered Alumina (93% Al_2O_3) had much higher levels of nitrogen ranging from 50% to 80% and a much lower level of CO_2 at 20% to 35% CO_2.

Therefore, depending upon the type of refractory material, a comparison can be made on different types of refractories.

Glass bubble analyses indicate that most of the origins of bubble defects are refractory related. These bubble compositions generally include air residuals (which corresponds to an Argon content within the bubble that matches the corresponding amount of Argon within air. Hence, the bubbles are usually the result of an opening of the refractory pores, whereby there is continuous corrosion. This is essentially continuous erosion along the glass sidewall contact blocks and the joints between these respective blocks.

Refractory cracks can also contribute to these gas bubbles. Essentially during either the heating up or filling, or thermal shocks during the furnace operation, these refractory cracks can contribute to refractory problems that continuously generate gas bubble defects.

There can also be penetration of the glass melt through to the sidewall block that comes in contact with the insulation layers. As a result there is significant contact with air from the insulation layer.

The high CO_2 content within the bubble generally indicates a longer residence time within the melt.

Sometimes the refractory bubbles associated with refractories, can possibly be confused with mechanically trapped air, mainly at low temperature zones. The refractory bubbles are usually of a small size and quite well dissipated within the glass. For example, these mechanically trapped air bubbles can come from changes in the glass level, from the operation of stirrers within the waist or forehearth areas, etc.

The other types of bubble compositions, but somewhat corresponding to refractory bubbles, are those from carbon impurities. As noted above, there are levels between 10% and 15% CO_2 in one group of refractories and 50% to 80% CO_2 in another. This can sometimes be confused with bubbles from carbon contamination (some contamination within the glass melt, or improper melting within the furnace). On the other hand, the bubble pressures for the melting problems are usually of a very low internal pressure, whereas the refractory problems are of a higher internal bubble pressure.

Generation of oxygen bubbles is generally an indication of a redox equilibrium process at the refractory surface, but normally at the beginning of the campaign. These oxygen bubbles have a very short residence time, and do not remain in the furnace very long. Oxygen bubbles can also be generated by a glass electrolysis mechanism. This reaction occurs due to an electrical interconnection of two conductive zones within the furnace, having different electrical potential, usually via furnace steel construction. The suspicious locations can be refractory cracks or open joints and subsequent hot glass penetration through the feeder or spout refractory blocks creating the cathode. The penetration of hot glass may touch the steel construction and close the electrochemical cell. Any failure in insulation from the steel construction of the boosting electrodes, coolers, batch charger, or thermocouple cladding creating the anode can cause the closing of the electric cell as well. Those options may start the electrochemical process.

In conclusion it can be seen that refractory material produces a general background level of gas bubble defects within glass, which can be affected by the initial and the resulting temperatures of the glass melt. To distinguish refractory bubbles from other types of bubbles, both the identity of the gases present within the bubble and the internal gas pressure of the bubble must be identified in order to obtain a proper diagnosis and solution.

ANALYSIS OF CORD AND STONES IN GLASS

Henry Dimmick, Neal Nichols, and Gary Smay
American Glass Research
Butler, PA and Maumee, OH

ABSTRACT
Glass containers are used successfully throughout the world for foods, beverages, spices, wines, liquors, etc. Depending on the type of product, glass containers can be subjected to a variety of loads such as internal pressure, thermal shock, impact or vertical load during normal filling operations and use. Each load causes deformations of the container which leads to the creation of stress in the glass. Usually, these stresses can be endured without any problems. However, on rare occasions, unusual conditions known as cord streaks and stone inclusions can be present in the glass which might adversely affect container performance. This paper will discuss how these conditions can be analyzed to find the source of potential problems.

INTRODUCTION

Glass is a very interesting material that can be used to create a variety of objects. Historically, glass has been used to fabricate ornate vessels for precious oils and ointments, art objects, stained-glass windows, decorative wine flasks, even tokens used in merchant exchange, etc [1,2]. Today, glass finds utility in hand-blown artwork, windows and containers. The focus of this paper will be on glass containers and some of the factors that can affect their performance during filling and use.

Soda-lime-silica glass is an almost ideal material for food and beverage containers. It is 100% recyclable with no memory of its prior use. After use, glass containers can be collected, crushed to an appropriate particle size and then re-melted. The use of large percentages of cullet can prolong the lifetime of a melting furnace and will result in operational fuel savings [3]. Glass has great consumer appeal and is perceived as a quality packaging material. Also related to consumer appeal is the ability of glass to be formed into a variety of designs and to be produced in a large number of colors. Soda-lime-silica glass is relatively inert, does not contain any hazardous constituents and will not leach any harmful components into the food or liquid it contains. Glass prohibits product tampering thereby providing a high level of safety and product integrity. Finally glass as a material is quite strong (up to 1,000,000 psi tensile strength [4]) and containers made from glass are able to withstand typical load levels during filling and use.

During use, glass containers typically endure internal pressure loads of up to 125 psi (8.6 bar) and on occasion pressures as high as 190 psi (13.1 bar). In the case of vertical load, it is common for containers to withstand 800-1000 lbs (365-454 kg) of vertical force during certain capping operations and typical thermal shock differentials can be as high as 50°F (28°C). Finally, glass containers are routinely subjected to impacts of 60-100 g-forces during filling operations. These load magnitudes are easily withstood with failure rates well less than 0.005 %. Appropriately, glass is considered as a superior packaging material.

The high success rate of glass containers is related to the criterion that controls glass failure. Any glass object fails when the tensile stress exceeds the glass surface strength at any specific location [4].

$$\text{Tension} \geq \text{Strength} \tag{1}$$

Usually, glass surface strengths are high and tensile stress magnitudes are low. Thus, glass containers have a very high survival rate. It is only when either the tensile stress magnitude is unusually high or the glass strength is abnormally low that problems arise. It is this condition that will be addressed in this paper relative to the presence of cord streaks and stone inclusions in the glass.

CORD AND STONE PROBLEMS

Cord
Cord is a condition that originates from non-homogeneous glass passing through the melting furnace. The primary source of this condition is the batching operation. Among the various factors that can cause such conditions is the placement of raw materials into the wrong storage bin, improper weighing of raw materials in the batch formulation, inadequate batch mixing, or batch segregation during storage and subsequent transport to the melting furnace [5,6].

These inhomogeneous streaks of glass have a composition that is different than the surrounding glass matrix. Associated with this compositional difference is a difference in the coefficient of thermal expansion (COE). As the glass container cools through the annealing lehr, the glass in the cord streak contracts at a different rate and amount compared to the surrounding glass matrix. Once the glass reaches room temperature, this difference in COE will result in the creation of stress in the glass. For cord streaks with a COE that is larger than the glass matrix, tensile stresses will be generated in the cord and compressive stresses will be induced in the surrounding glass. For cord streaks with a COE that is smaller than the glass matrix, compressive stresses will be generated in the cord and tensile stresses will be induced in the surrounding glass.

Since glass only fails under the action of tensile stresses [7], cord streaks that directly create a tensile stress and those tensile stresses induced by compressive cord streaks will be of concern. The magnitude of these tensile stresses will mathematically add to tensile stresses generated by applied loads. The net effect is that the total tensile stress in equation 1 will be larger than anticipated from the applied loads only. Consequently, bottles with cord streaks exhibiting high enough tensile stress can fail at loads that are lower than typical levels. The performance of the container is subsequently adversely affected.

Stones
Stones are un-melted or re-crystallized solid materials in the amorphous glass matrix. Stones can originate from a variety of sources among which are batch segregation or excessive amounts of sand in the batch, cullet contaminants such as oven ware, ceramic materials or aluminum metal, erosion of furnace refractories either above or below the metal line that do not subsequently melt and diffuse into the molten glass and devitrification, a condition in which the molten glass is maintained for some extended time below its liquidus temperature, resulting in nucleation and crystalline growth in the glass [5].

Most stones have a certain COE associated with them. If the COE differs significantly compared to the glass matrix, large stresses will be generated in the surrounding glass. For those situations where these stresses are tensile in nature, they will mathematically add to the tensile stresses generated by applied loads. The net effect is that the total tensile stress in equation 1 will again be larger than typically anticipated from applied loads only. In other situations, the stones will act as flaws in the glass decreasing glass strength. This can be related to the surface morphology of the stones or in some instances, tensile stresses created by differences in the COE will cause microcracks to form in the glass surrounding the stone. The net effect is that the glass strength in equation 1 will be much smaller than typically anticipated. Consequently, bottles containing stones can fail at loads that are lower than typical levels either due to lower strengths or higher stresses.

For both cord streaks and stone inclusions, it is important that a complete analysis be undertaken to determine the source of the potential problems. This will enable corrective action to be taken for the immediate problem. In addition, proper analyses will indicate how to avoid the problem in future production runs.

ANALYSIS OF CORD AND STONES

Cord

Typical cord streaks are usually aligned with the vertical axis of the bottles as shown in Figure 1a. In order to determine the stresses that are present in these cord streaks, ring sections are routinely prepared from the sidewall of the test container as shown in Figure 1b. These ring sections are analyzed in a polarizing microscope according to the procedure detailed in ASTM Standard Method C-978. In this procedure, polarized light is passed through the ring section as indicated schematically by the arrow in Figure 1b where it interacts with the stresses in the cord. Examples of ring sections containing cord streaks as viewed in a polarizing microscope are shown in Figure 1c. In these views, the polarized light is aligned perpendicular to the plane of the paper.

The color observed in the polarizing microscope is the retardation of polarized light created by the cord streaks. This value, R, is related to the stresses in the cord streak, S, the optical path length, d, and stress optical coefficient of the glass, C, according to the equation:

$$R = SCd \tag{2}$$

The retardation value can be quantitatively measured with various types of compensators such as a Berek tilting compensator or a Senarmont rotating compensator attached to a polarizing microscope. The measured retardation value, R in nm, is converted to stress, S in kg/cm^2, by the following relationship [8]:

$$S = 0.38 \ (R/d) \tag{3}$$

where d, in cm, is the optical path length (the thickness of the ring section). The proportionality factor 0.38 takes into account the units of retardation and the units of the stress optical coefficient for soda-lime-glass. Limits relative to the allowable magnitudes of cord stresses are determined by individual companies and are typically related to the specific end-use of the containers.

While measurement of cord stresses are important, it provides only limited clues as to the source of the cord. Therefore, it is important to determine the composition of the cord in order to specifically identify its source. Such analyses can be done with the use of focused beam instrumentation such as an energy dispersive X-ray spectrometer (EDX) in conjunction with a scanning electron microscope (SEM) or a wavelength dispersive X-ray spectrometer (WDX) in conjunction with an electron microprobe. These techniques can be used to analyze discrete areas of cordy glass and compare those results with compositional analyses of the surrounding glass matrix. Analyses by EDX provide only near quantitative data. Thus, a precise determination of the composition of the cord cannot be obtained using this technique. However, under careful operating procedures, it is possible to obtain a relative composition of the cord compared to the glass matrix. Determining those oxides that are enriched in the cord compared to the glass matrix will often provide sufficient information to identify the source of the cord. More accurate information about the composition of the cord streak can be obtained by the use of a wavelength dispersive X-ray spectrometer (WDX) as it is much more sensitive to the lighter elements on the periodic chart compared to the EDX technique.

A test procedure for either EDX or WDX analyses is shown in Figure 2. In this process, a series of line analyses are performed at numerous locations, approximately 5-7 microns apart, traversing from the glass matrix, across the cord streak and back into the glass matrix region. Each individual line analysis is approximately 200 microns long. The electron beam parameters are held constant with the beam slightly defocused to minimize the loss of the sodium and magnesium signal. Typical analyses consist of sodium, magnesium, aluminum, silicon, potassium, calcium, zirconium and iron. Regions in the glass matrix are averaged together and the signal is standardized to the actual composition of the glass as obtained by other quantitative techniques. Regions in the cord streak are averaged together and then compared to the glass matrix. This comparison provides information about the specific composition of the cord relative to the glass matrix.

Thus, either EDX of WDX can be advantageously used to obtain information about the composition of the cord streak. It should be noted that cord exhibiting stresses less than about 28 kg/cm^2 have a very minor compositional difference compared to the surrounding glass. This minor compositional difference cannot be detected using these methods. However, cord stresses less than 28 kg/cm^2 are not problematic to the overall performance of containers, are typically not of concern and would not need to be analyzed [9].

Stones

When stones occur, they are found randomly distributed throughout the thickness of the glass. Analyses of these stones require that the glass be fractured or ground-down in order to expose the stone. Typical analyses are then done according to accepted industry practices using the physical characteristics or crystalline properties of the stones.

Physical characteristics of the stones such as topography, color, presence of stress in the glass, indications of melting into the glass, etc. can sometimes be used to identify the source of the stone. Two examples will be presented to show this process.

- The stone in Figure 3 is smooth in nature, white in color and exhibits multiple, small, rounded granular features. This is often termed a "grape cluster" stone due to its similarity in appearance to a cluster of grapes. These characteristics are unique to a silica batch stone consisting of small rounded grains of unmolten sand. This stone occurs when the sand grains have not been exposed to soda in the melting furnace. These stones are due to excessive sand in the batch or to batch segregation during mixing or storage.
- The stone in Figure 4 consists of what is termed a solution sack (cordy knot) containing distinct crystals with the appearance of snow flakes. The combination of the solution sack with the crystals are typical for AZS refractories which have partially or fully melted with a subsequent recrystallization of either zircon or zirconia. These stones are usually the result of erosion of the furnace refractories.

Crystalline properties can be determined by creating a thin section of the stone and subjecting it to standard petrographic analyses using a polarizing microscope. Two examples of this procedure will be presented to show this process.

- The stone in Figure 5 shows unique crystals that are rod-like in appearance. As a thin section, this stone exhibits a very high degree of birefringence. These characteristics are uniquely typical of beta-wollastonite, a common devitrification stone that occurs when a standard soda-lime-silica composition is allowed to remain at temperatures below the liquidus point.
- The stone in Figure 6 consists of a flaky, metallic appearance. In thin section, the crystals in the stone consist of lathes that are random in orientation. These characteristics are uniquely typical of chromic oxide stones that originate from unmolten colorant or as intrinsic contaminants of sand.

Finally, EDX or WDX techniques are often useful to determine the compositional details of stones. A few examples will be provided of this type of analysis.

- The stone in Figure 7 is bluish in color, smooth and rounded in appearance. When examined in a scanning electron microscope (SEM), the cross-sectional view in Figure 7b is obtained. EDX analyses indicated that the stone was composed solely of alumina (aluminum and oxygen). Based on the physical appearance and EDX composition, it was concluded this stone was corundum. One source of this type of stone is from cullet contaminated with the remnants of grinding wheels.

- The stone in Figure 8 is smooth and spherical in appearance with a few reaction bubbles around the perimeter. Viewed in cross-section in the SEM shows a very consistent appearance with a few light colored crystals. EDX analyses indicated that the bulk of the stone was composed of silicon and that the light colored segments were composed of titanium, iron and copper. Based on the physical appearance and the EDX composition, it was concluded this stone was a silicon ball. It is created from the oxidation/reduction reaction of aluminum metal with molten silica in the melting furnace. The metallic aluminum is oxidized into alumina which diffuses into the surrounding glass. The molten silica is reduced into elemental silicon which forms the spherical stone. The small crystals in the silicon ball are the remnants of titanium, iron and copper which often are present in aluminum metal caps. Thus, this stone most likely originated from aluminum caps in the cullet which was used in the batching operation.

EXAMPLES OF CORD AND STONE ANALYSES

Case Study #1 (Cord Issue)

During routine examination of ring sections, multiple cord streaks were observed some of which were buried in the interior of the glass while others were located on the outside surface of the bottle. First considerations involved the stress level of these cord streaks and whether they may be excessively high. Examination of the ring sections in a polarizing microscope revealed that the cord streaks were tensile in nature and that the stress magnitude of the streak on the outside bottle surface was excessively high. Such high stresses could adversely affect the performance of the bottles relative to internal pressure or thermal shock resistance. Thus, the ware was discarded at the production plant.

Further analyses involved a determination of the source of the cord streaks. Based on the creation of tensile stresses in the cord, it was concluded that the streaks were enriched in either soda or calcia, oxides that have a COE that is higher than the glass matrix. The ring section containing the highest stressed cord on the outside bottle surface was polished to provide a flat surface for EDX analyses. The composition of the cord was determined as an average of ten measurements in the center of the cord streak. These results were compared to a similar ten measurement average in the glass matrix as shown in Figure 9 and Table I. These data indicated that the tensile cord was enriched in calcia compared to the matrix glass.

These results provided the necessary information to pursue problems specifically related to the limestone raw material. It was subsequently found that a load of limestone was inadvertently dumped into the silo containing sand. This caused excessive amounts of limestone to be included into the batch which resulted in poor mixing. This produced an inhomogeniety in the furnace and ultimately created a tensile cord streak in the bottles. The solution was to carefully monitor the arrival of the raw materials so that they are placed into the correct storage silos.

Case Study #2 (Stone issue)

A container was exhibiting excessive levels of breakage from an internal pressure load. Examination of the broken bottles indicated that the pressure load was normal. Further examination revealed the presence of stones at the fracture origins. These stones exhibited a very smooth, metallic shine and were perfectly spherical, as shown by the example in Figure 8. Examination in a polarizing microscope indicated that substantial stresses were present in the glass surrounding the stone. Based on the physical characteristics of the stones, the initial conclusion was they were silicon balls.

The supposition that the stones were composed of silicon was confirmed by EDX analyses. It is imperative in the use of cullet that all metals, especially aluminum, be avoided to assure that this problem is eliminated in the future.

These examples are indicative of the type of analyses that should be employed to determine the primary cause of cord streaks and stone inclusions. Through a systematic approach to these problems, the source of the cords and stones can be determined and corrective actions can be implemented. Careful record keeping can also be used to help avoid these types of problems in future production runs.

REFERENCES

1. "History of Glass", Wikipedia

2. Anonymous, "A Historical Look at Glass, The History, It's Nature, and it's Recipe, Articles & Links on the History and Properties of Glass", courtesy of PPG Industries, Inc.

3. Glass Melting Technology: A Technical and Economic Assessment, Glass Manufacturing Industry Council, October 2004.

4. R.E. Mould, "The Strength of Inorganic Glasses", Fundamental Phenomena in the Materials Sciences, Vol 4, 1967, p119

5. Stones and Cord in Glass, C. Clark-Monks and J. M. Parker, Society of Glass Technology, 1980.

6. M. A. Knight, "Cords in Glass", The Glass Industry, September-December, 1956.

7. Fractography of Glass, Fracture of Glass Containers, John B. Kepple and John S. Wasylyk, Edited by R. C. Bradt and R. E. Tressler, Plenum Press, New York, New York, 1994

8. "Photoelastic Determination of Residual Stress in a Transparent Glass Matrix Using a Polarizing Microscope and Optical Retardation Compensation Procedures", ASTM Standard Test Method C978.

9. Unpublished results

"CAT SCRATCH" CORD DISPERSAL

Les Gaskell
General Manager – Technical Services Division
Parkinson-Spencer Refractories Limited
Holmfield, Halifax, West Yorkshire, HX 3 6SX, United Kingdom

ABSTRACT

"Cat scratch" cord has been a significant problem in the glass industry for perhaps seventy years. It is a common problem in tableware and containers but can also be present in rolled plate flat glass and tubing. "Cat scratch" cord is caused by a viscous glass enriched in alumina and zirconia which settles out on the bottom of furnaces, distributors and forehearths. This glass eventually travels along the bottom of the forehearths, is present in the surface of the gob and appears as a line or series of lines in the surface of the bottle or article being manufactured. One of the most effective ways of dealing with this defect is by the operation of correctly configured stirrers to lift and disperse this viscous glass into the body of the base glass so that it is then present in the centre of the gob and distributed throughout the base and side walls of the article so that it is no longer readily visible. This paper describes the application of stirrers in a proven cord dispersal system. It is based on the observations and experiences of Parkinson-Spencer Refractories (PSR) in supplying refractory stirrers and complete cord dispersal stirrer systems to the glass industry worldwide.

CORD

Cord is a vitreous defect in a glass product which has a different composition and hence physical properties from the base glass being produced. It can be caused by problems with the composition of the raw materials, incorrect weighing or mixing of the raw materials during batch preparation, segregation of the batch during transport to or charging into the furnace, inadequate homogenisation during the melting process and the loss of volatile components of the glass from the glass surface during melting and conditioning. It can also be caused by the dissolution of refractories into the glass.

"CAT SCRATCH" CORD

"Cat scratch" cord is a surface cord originating from the dissolution of refractories. It exhibits itself as a line or series of lines on the surface of the bottle or article being manufactured. These lines have the appearance of a cat having scratched down the side of the product with its claws, hence the most popular name of "cat scratch" cord for this defect. Other less commonly used names to describe this defect are "cat's whiskers", "angel's hair", "mare's tail", "surface cord", "refractory cord", "feeder marks" and "zircon cord".

Virtually all tableware and container manufacturers have this problem to varying degrees. It can be a particular problem in thin blown tableware as the surface cord contributes to a greater percentage of the overall glass thickness resulting in a highly visible defect unacceptable for good quality tableware. It is not usually a problem in pressed tableware due to the greater glass thickness. It is a visual defect in normal quality glass bottles which can usually be tolerated if limited in extent but can become unacceptable if it becomes extreme or is present in premium quality bottles. It can also occur in other glass products such as rolled plate glass and tubing.

The "cat scratch" cord defect has the following characteristics:

- A line or series of lines on the container surface.
- It is considered a visual defect not significantly affecting the container strength.
- It can very often be felt with the finger nail as ridges on the glass surface. These ridges can sometimes cause problems with any subsequent labelling or other surface decoration of the article.
- The cord is normally associated with a low compressive stress.
- The surface cord is enriched in alumina and/or zirconia compared with the base glass by typically 2 to 6%. As the defect is on or very close to the surface it is very difficult to analyse even by modern sophisticated techniques.
- Due to the enrichment in alumina and/or zirconia the surface cord material is more viscous than the base glass.
- The glass on the surface of the article is formed by the glass on the surface of the gob which is from the bottom glass in the feeder spout and forehearth.
- As the cord material is more viscous than the base glass it tends to settle out and build up on the bottom of forehearths and feeder spouts.
- On multiple gob feeder operation with centre line shearing it normally only occurs or is much worse on the gob closest to the forehearth which is the gob receiving the bottom glass first.
- On multiple gob feeder operation with 90 degree shearing it can occur in all gobs but the gob on the side to which the feeder tube rotates is normally the worst which is the gob receiving the bottom glass first.

Typical "cat scratch" cord in a white flint glass bottle.

- It is more likely to occur on large containers manufactured with low gob temperatures. The viscous cord material is more likely to settle out at lower temperatures.
- It is more likely to occur on the outer forehearths of a multiple forehearth distributor layout. The viscous cord material is more likely to settle out due to the longer residence times to the outer forehearths.
- The severity of the cord can change and it can come and go with changes in individual forehearth, distributor section and overall furnace pull and operating temperatures.
- Once present in a particular product with the forehearth operating under steady state conditions it is normally consistently positioned and present in every gob and in every bottle produced by that gob.
- During periods when the cord material is building up or dispersing its position can change and it can be present intermittently in the gob.
- It can occur on new furnaces at the very start of a furnace campaign, can be present throughout an entire campaign, can come and go during a campaign, as well as appearing on old furnaces towards the end of a campaign.

- It has been a widespread and persistent problem over perhaps 70 years indicating that it is fundamentally linked to the glass melting process and the refractories normally used.

THE SOURCE OF "CAT SCRATCH" CORD

Based on the characteristics of "cat scratch" cord the most probable source of the viscous cord material is the exudation on initial heating and/or subsequent corrosion of the vitreous (glassy) phase of the fusion cast alumina-zirconia-silica (AZS) refractories used in the furnace melter tank and superstructure. These refractories typically contain around 20% vitreous phase. Up to 3% of this vitreous phase can exude from the surface of these refractories on initial heating. Dissolution of the vitreous phase also occurs during the normal subsequent corrosion of the refractory. If fusion cast AZS refractories are used in the distributor and forehearth they could also be a possible source of "cat scratch" cord but the corrosion rate is much lower than in the melter.

When occurring at the very start of a furnace campaign, within days of the start-up, "cat scratch" cord must be associated with the exudation of the vitreous phase from the AZS refractories which takes place during the initial furnace warm-up as no significant corrosion of any refractories can have taken place at that time. When occurring on old furnaces it is associated with the dissolution of the vitreous phase which takes place during normal subsequent corrosion of the fusion cast AZS refractories.

The cause of "cat scratch" cord has been a source of controversy amongst glass manufacturers and refractory suppliers for many years. There is a longstanding belief among many glass manufacturers that the source of the "cat scratch" cord material is predominantly zircon or zirconia containing refractories when used downstream from the furnace throat, hence the name "zircon cord". However, glass companies that use zirconia-free refractories such as fusion cast alpha-beta alumina glass contact blocks in the distributors and forehearths can still suffer from "cat scratch" cord. Conversely, glass companies using fusion cast AZS refractories in the distributor and bonded AZS refractories in the feeder and forehearths do not always experience problems with "cat scratch" cord.

Whilst the fusion cast alpha-beta alumina refractories recommended by fusion cast refractory suppliers for glass contact use beyond the throat in distributors and forehearths contain no zirconia, they also contain minimal glassy phase and do not exhibit exudation on initial heating. Bonded AZS products used for feeder expendables and forehearth channels and bonded alumina products used in the distributor and forehearth also contain no glassy phase and should not normally be a source of "cat scratch" cord.

When a glass furnace is examined at the end of a campaign a considerable amount of refractory material is missing having dissolved into the glass. It is a testament to the development and quality of the fusion cast AZS refractories used that under normal circumstances this material is assimilated into the melt, remains dispersed and only registers as minor changes to the final glass analysis. However, if the corrosion rate is excessive at any time during the campaign due to a particular operating problem, or the corrosion products build up in an area of the furnace, distributor or forehearth and then subsequently become mobile, this can result in the corrosion products being concentrated rather than dispersed in the glass and then visible as the "cat scratch" cord defect.

"Cat scratch" cord appears to have become more common with the introduction of distributors to replace conventional refiners as well as with the increase in the specific output of furnaces. In the past refiners were designed as part of the furnace and were sized relative to the furnace melter. Generally, the refiner would be designed with a surface area about one third to a quarter of the melter and with a similar glass depth. Such oversized refiners would have virtually stagnant glass on the refiner bottom which would provide a natural reservoir to trap the viscous corrosion products from the melter. Indeed, very often devitrified glass would be found at the bottom of these oversized refiners during furnace rebuilds. Refiners have gradually been reduced in size and depth over the years and evolved into the modern distributor.

The modern distributor is much smaller in area and shallower in glass depth than the old conventional refiner to allow much better control of the glass temperature of the individual glass streams to each forehearth. However, the modern distributor provides no reservoir to trap the corrosion products from the fusion cast refractories in the melter which continually pass through the distributor in the glass and can, under suitable conditions, settle out and manifest themselves as " cat scratch" cord.

ACTIONS TO REDUCE THE INCIDENCE OF "CAT SCRATCH" CORD

Whilst the source of the problem cannot be eliminated with the current refractories available the following preventative actions can help to reduce the incidence of "cat scratch" cord:

- Furnaces should be designed and operated to minimise excessive refractory corrosion.
- The temperature to which furnaces are heated during the warm-up before commencing the filling-on of the furnace with cullet should be limited to minimise exudation from the AZS refractories (both glass contact and superstructure) and the associated build up of vitreous phase on the furnace bottom. If the furnace is heated up to the maximum operating temperature and held at this temperature empty for a period of time before filling-on the furnace with cullet, the exuded glassy phase built up on the furnace bottom will be pushed through the throat during the subsequent filling-on of the furnace.
- Furnaces should be designed and operated to maximise convection currents within the glass to homogenise the glass and prevent viscous corrosion products from settling out.
- Furnaces should be designed to prevent the build up of cold stagnant glass where the viscous corrosion products can settle out.
- Periods when the furnace is operated at low loads which could allow viscous corrosion products to settle out should be avoided or minimised.
- Furnaces should be designed and operated to minimise batch carry-over which can cause excessive corrosion of the superstructure refractories and run down into the melt.
- Flame lengths in furnaces should be controlled to prevent flame impingement on breast walls and port openings which can cause excessive temperatures, corrosion and run down of corrosion products into the melt.
- Distributors and forehearths should be designed and operated to avoid areas in which corrosion products can settle out and accumulate.
- If fusion cast AZS refractories are used in the distributor and forehearths then the temperature to which they are heated up before being filled with glass should be limited to minimise the exudation and prevent the build up of the vitreous phase on the bottom of the distributor and forehearths.
- The "cat scratch" cord material will settle out and build up in deeper and colder areas of the distributor and forehearths, in forehearths operating with low glass temperatures, poor glass thermal homogeneity and long glass residence times. Such areas and conditions should be avoided by correct design and operation.
- The "cat scratch" cord material will build up in any defects in the forehearth and feeder refractories such as cracks and corroded areas. Such areas should be prevented by correct installation and warm-up procedures and repaired when possible.
- Colourant forehearths have deeper colourant sections (frit addition and stirrer sections) in which the "cat scratch" cord material can settle out during white flint operation at lower operating temperatures and then become mobile with the action of the stirrers at higher operating temperatures. Colourant forehearth sections also often use fusion cast AZS glass

contact and superstructure refractories which can be a further source of the "cat scratch" cord material because of the much higher wear rates due to the volatile materials present from the frit and the higher operating temperatures during colourant operation. Colourant forehearths should therefore be equipped with systems to eliminate or minimise any possible "cat scratch" cord defect.

METHODS TO ELIMINATE THE "CAT SCRATCH" CORD DEFECT

Other than temporary, short term methods to eliminate "cat scratch" cord such as changing the direction of the feeder tube rotation, replacing the feeder spout or draining the forehearth of glass, the two main methods adopted are continuous draining and stirring.

Draining relies on collecting and removing all the corrosion products which have settled out before they reach the feeder spout. It requires a special drain block or sump to be provided where the "cat scratch" cord material can be collected and then drained out of the glass flow. Drains were initially installed in the bottom of the furnace throat but this was not entirely successful because the viscous corrosion products do not all settle out at this point and are more likely to settle out in the forehearths and distributor due to the much lower temperatures and throughput. They are now installed near the end of each forehearth to be sure to collect the material wherever it may settle out during the conditioning process. The major disadvantages of the drains are the loss of glass and potential production and the fact that a special drain block needs to be installed to collect the "cat scratch" cord material. Whilst the special drain blocks can be installed during a furnace rebuild in case of a future potential "cat scratch" cord problem they can also provoke a problem by collecting the material making it then absolutely necessary to drain the glass.

Stirring relies on re-dispersing the corrosion products into the glass before they reach the feeder spout and has proved to be one of the most effective ways of dealing with the visual "cat scratch" cord defect. The stirrers lift the viscous glass off the bottom of the forehearth channel and mix it into the body of the glass so that it is no longer concentrated on the gob surface and is distributed throughout the base and side walls of the article during the forming process so that it is no longer visible in the finished product. This requires the stirrers to be correctly designed, located and configured.

STIRRER CONFIGURATION FOR "CAT SCRATCH" CORD DISPERSAL

To ensure success the stirrers must be located in the equalising section of the forehearth. This ensures that any material which may be settling out along the forehearth is dispersed and the material cannot settle out again before it reaches the feeder spout.

In the past helical stirrers have been used as blenders in forehearth equalising sections to improve the glass thermal homogeneity at the spout entrance. They are positioned near the channel side walls and are configured to lift the bottom glass and move glass from the channel side walls to the centre of the forehearth. As viewed from the spout against the direction of glass flow, the stirrer on the left-hand side should be a right-hand helix stirrer rotating clockwise and the stirrer on the right-hand side should be a left-hand helix stirrer rotating anti-clockwise. This stirrer configuration is designed to lift the bottom glass and push the glass at the side walls towards the centre of the forehearth and spout. Traditional equalising section blenders are not suitable for the dispersal of "cat scratch" cord because the viscous glass can pass between the stirrers, particularly in wider equalising sections.

However, twin counter-rotating helical stirrers can be used successfully in narrow equalising sections of 16, 22 and in some circumstances 26 inches width for the dispersal of "cat scratch" cord. They have the same handing and rotation direction as blenders but must be operated as close to the channel base as possible and correctly spaced across the channel width.

For wider equalising sections of 26, 36, 43 and 48 inches width twin counter-rotating paddle type stirrers must be used to provide adequate coverage of the channel base. The paddles have left-hand and right-hand blades and are rotated in the same configuration as the helical stirrers. They must be operated as close to the channel base as possible. The paddles are operated at 90 degrees to each other and are normally configured to overlap to provide improved stirring efficiency.

Twin counter-rotating helical or paddle stirrers are used rather than a single paddle stirrer because they direct the glass away from both channel side walls and towards the centre of the spout so that the feeder tube can be rotated in either direction as required by the forming personnel. If a single paddle stirrer is used it will always direct glass to one side of the forehearth. It needs to be configured to rotate in the opposite direction to the feeder tube and the feeder tube rotation direction then needs to be fixed. If the stirrer rotates in the same direction as the feeder tube the "cat scratch" cord can actually be intensified.

STIRRER INSTALLATION

An access slot approximately 9 inches long and the full width of the channel block with its centre line typically 36 inches back from the spout entrance is required in the superstructure refractories for the installation of the equalising section stirrers. Stirrer cover tiles are required for sealing around the stirrer shafts when the stirrers are operating. Stirrer position cover tiles are required for sealing the access slot when the stirrers are not in use.

Unless the existing equalising section superstructure has been designed with a suitable access slot for the installation of the stirrers, the equalising section superstructure refractories have to be replaced to accommodate the installation of the stirrers. This can be done hot during a reasonably long job change or other machine stoppage such as during a spout change.

The superstructure refractories are supplied in a material suitable for hot insertion. Suitable roof block lifting frames and temporary lightweight roof block covers are supplied to assist with the hot replacement operation. The replacement equalising section superstructure refractories are supplied to match as closely as possible the existing design whilst incorporating the access slot required for the stirrers. This allows the possibility for the minimum amount of refractories to be replaced under hot conditions to minimise production downtime. The stirrer mechanism and support frames can then be installed on-the-run during the normal operation of the forehearth once the superstructure refractories have been replaced.

STIRRER MECHANISM

The stirrer mechanism comprises the following:

- Mounting brackets for mounting the support frames to the equalising section casing.
  The stirrer mechanism is normally mounted to the casing top flange. However, the stirrer mechanism can be designed for support steelwork or forehearth platform mounting if the existing forehearth casing support is considered inadequate.
- Support and lifting frames.
  The stirrer mechanism support and lifting frame allows the mechanism to be raised above, pulled out to the side of the forehearth and then lowered to forehearth platform level to allow easy access for stirrer replacement and mechanism maintenance. The stirrer mechanism mounting brackets and support frame are designed to fit the existing forehearth superstructure bracing steelwork, burner pipe work and feeder mechanism as closely as possible but some slight on site modifications to the bracing steelwork and pipe work may still be necessary.
- Stirrer back plate assembly with air cooled carbon bearings and stainless steel chucks.
  The stirrer mechanism back plate assembly uses air-cooled carbon bearings and stainless steel stirrer chucks for accurate, concentric stirrer rotation, reliability of operation and minimal maintenance requirements. Adjustable shaft mounted bevel gears are used to enable reliable setting of the required relative stirrer positions during installation.

- Variable speed drive motor and gear box.
  The stirrer variable speed drive motor and gear box are mounted on the stirrer cross beam for easy access and are provided with a cable management system to allow the stirrers to be operated outboard at the side of the forehearth, above the forehearth or with the stirrers above the glass in the forehearth combustion space to check for correct stirrer operation before installing the stirrers into their final operating position in the glass.
- Drive motor speed control panel.
  The stirrer system control panel houses the inverter drive which controls the stirrer drive motor speed and provides indication and control of the stirrer rotational speed as well as local and remote alarming of stirrer faults. Access to the speed control and inverter drive is locked to prevent changes by unauthorised personnel. The control panel is normally mounted in a control room but can be supplied equipped with an integral air-conditioning unit for mounting on the machine shop floor.

Illustration of typical equalising section stirrer mechanism.

To design the stirrer mechanism to fit the particular installation the following information is required:

- Drawings showing details of the existing equalising section substructure and superstructure refractory assembly.
- Drawings showing details of the existing equalising section casing and support steelwork.
- Drawings showing details of the existing equalising section combustion system burner manifolds and pipe work.
- Maximum available headroom for the stirrer mechanism support frame.
- The side of the forehearth to which it is more convenient for the stirrers to be pulled out for stirrer replacement and mechanism maintenance.
- Details of the factory three phase and single phase electrical power supply.
- Photographs showing existing equipment installed in the vicinity of the equalising section.

STIRRER OPERATION

The stirrers should be operated as slowly as possible to achieve the dispersal of the "cat scratch" cord with minimum stirrer and additional channel block refractory wear. Stirrer rotational speeds are normally in the range of 2 to 15 revolutions per minute with 5 revolutions per minute being typical.

Initially any cord material built up downstream from the stirrers on the equalising section channel blocks and spout must be "washed out" with the "clean" glass provided from the stirrers. Carrying out the stirrer system installation during a spout change helps to speed up this process as the equalising section is normally drained of glass and the spout is new.

Following this initial period the "cat scratch" cord appearance can be closely and predictably controlled by adjusting the stirrer speed. The stirrer speed can be reduced to a minimum as the cord condition improves.

The stirrers are normally continuously operated at a minimum speed to guard against the appearance of "cat scratch" cord. However, they can be stopped and raised above the glass or completely removed if considered necessary when no cord is present.

The maximum stirrer speed possible depends upon the particular equalising section temperature required for forming. If the stirrers are rotated too quickly at a low glass temperature the high glass viscosity and resulting high shear stresses between the glass and the refractory may result in blister formation.

The stirrer operation does not interfere with thermocouple readings or gob weight control and can help to improve the thermal homogeneity of the glass at the spout entrance. It can, however, result in some fluctuation in the reading of any adjacent downstream infra-red thermometer due to the disturbance of the glass surface but this can normally be overcome by filtering of the temperature signal.

Stirrer operation does not increase glass head loss and can help to minimise glass head loss at high forehearth loads.

Stirrer life depends on the glass composition, glass temperature and stirrer rotational speed. Typical stirrer life would be 3 to 9 months. As the refractory stirrers wear over time the stirrer rotational speed needs to be increased to obtain the same stirring effect.

Increased channel block wear at the stirrer position is minimal over a furnace campaign. Any local additional long term wear can be accommodated by stirrer operating height adjustment. This can in fact be beneficial to the stirring operation as the stirrers then operate at a height lower than the adjacent upstream channel block base to better lift the cord.

Stirrer system in operation.

SUMMARY

Correctly designed, installed and operated stirrer systems have been found to be a guaranteed method of eliminating or reducing to an acceptable level the visible "cat scratch" cord defect. They have advantages over alternative methods such as drains in that there is no cost of the waste glass and lost potential production. The stirrer systems can also contribute to improving the glass thermal homogeneity at the spout entrance. PSR has supplied nearly 50 Cord Dispersal Stirrer Systems to date all with complete success.

TOOLS USED TO IMPROVE OPERATIONAL SAFETY IN JOHNS MANVILLE GLASS PLANTS

Noel Camp
Johns Manville

ABSTRACT

Johns Manville (JM) has a long-standing core value of safety at its operations around the world. As focused as the company has been on ensuring a safe work environment for its people, in the mid-1990s the need to balance regulatory compliance, employee engagement, and risk elimination led to a 15-year effort to modernize safety processes. This change process resulted in a safety program delivering a sustainable step-change in improved safety performance. Ultimately, the changes resulted in the development and refinement of formalized safety tools and approaches proven effective at identifying risk, and enabling the company to modify designs and reduce the potential for incident occurrence. Three of the most effective tools in managing risk associated with glass manufacturing operations include Management of Change (MOC), Process Hazard Analysis (PHA), and Technical Inquiry (TI). The MOC process provides a structured approach to evaluating many dimensions of potential risk associated with critical processes prior to making a physical change. In cases where the results of an MOC analysis suggest significant risk, a variety of PHA tools were developed to identify and correct probable failure modes in order to arrive at an acceptable risk profile. In situations where desired outcomes are not achieved and significant safety incidents have occurred, the TI process is employed to maximize learning and minimize the potential for a repeat of a loss event by effectively identifying and correcting root causes.

A BALANCED APPROACH TO SAFETY PROCESSES

Since entering the glass business in the mid-1950s Johns Manville has been involved in the manufacture of quality products for a variety of markets including roofing materials and fiber glass products for building materials, aerospace, automotive and filtration applications. Like many companies, the JM operating philosophy regarding safety evolved and improved over time to place greater emphasis on worker protection. With improved awareness of the correlation between workplace exposures and the financial condition of an organization, JM initially placed greater emphasis on regulatory program compliance. Through enhanced focus on regulatory program compliance, JM lowered Total Case Injury Rate (TCIR – on an annual basis, the number of employees per 100 that require medical treatment beyond first aid to treat a work related injury or illness) to approximately 8.5 during the early to mid 1990s.

During this time JM's safety management system utilized thorough regulatory audits, completed at each factory every three years. JM enhanced this audit process with a safety management system featuring a 650-point Safety Management Rating (SMR), involving regulatory programs and execution of those programs. Use of the SMR as a primary component of the safety management system was largely ineffective in improving safety results as measured by TCIR. Despite the detailed SMR criteria, the process remained focused on minimal regulatory requirements such as electrical energy control features, confined space entry practices, and chemical hazard awareness. In the mid-1990s, several events provided further catalyst for change in Johns Manville's safety practices.

In 1996, a fatality occurred in a Johns Manville facility. An electrical engineer was electrocuted as the result of a change in operating procedures that had not been fully analyzed prior to making the change. In response to that event, JM's senior leader demanded an exceptionally thorough incident investigation. The investigation included extensive analysis and discussion and led to the conclusion that the almost exclusive focus on compliance with regulatory programs and processes was simply insufficient to prevent severe incidents.

Later in 1996, Johns Manville initiated a change toward a more balanced approach to employee safety. A new philosophy evolved toward a basic objective to ensure that employees go home in the same state of health as when they arrive at work. This change required identifying and mitigating hazards (whether required by regulation or not) prior to incident and proactively engaging employees in safety efforts.

The current Johns Manville Safety Management System (SMS) is built around the OSHA Voluntary Protection Program (VPP) model. Within the current system, safety efforts are broken down into four key areas: Management Leadership and Employee Involvement, Worksite Analysis, Hazard Prevention and Control, and Health and Safety Training. Notable among new additions to the safety efforts are several key tools focused on identifying and mitigating risk. These new tools (Management of Change (MOC), Process Hazard Analysis (PHA), and Technical Inquiry (TI)), have been critical to delivering a sustained 80% reduction in TCIR since 1995.

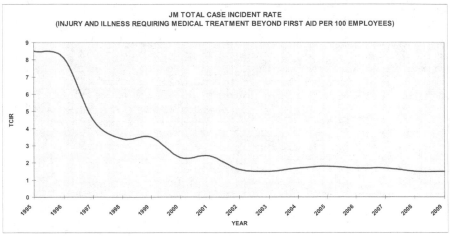

Figure 1: Johns Manville TCIR Progression

PROCESS HAZARD ANALYSIS

As an integral part of the increased emphasis on a truly balanced safety program, JM determined that injury trend analysis techniques required improvement. Initially, JM trend analysis identified very specific injury trends, such as employee wrist fractures that are common to facilities with a certain packaging technology. These types of observations resulted in a better understanding that although equipment and work instructions could be in compliance with regulatory requirements, the physical location of features such as machine guards and lockout points could still make the systems ineffective and increase incident potential. Relocating the machine guards or energy isolation points quickly resolved these specific injury trends. Once these easily identifiable injury features were resolved, JM found that the absence of further identifiable trends made additional improvement more challenging.

In 1999, company's insulation business sought to improve its manufacturing process design systems. An objective was to design manufacturing equipment that would be inherently safer to operate and maintain than existing JM assets. Investigation of safe manufacturing practices used in the hazardous chemicals industry provided valuable lessons that JM applied to its own safety practices. The investigation found that prior to finalizing equipment designs in facilities producing highly hazardous chemicals, teams would evaluate process hazards that may result in loss. Most often this evaluation is done specifically under requirements of the PSM regulation with which these types of operations must comply. Using this integrated design approach, hazards are engineered out of the systems.

JM initially applied Process Hazard Analysis during design and construction of a new light density fiber glass wool line in one of its insulation plants. JM used a consultant who was an expert in chemical processes to assist with this first use of the PHA tool. Unfortunately, this resulted in recommendations geared too much toward the chemical industry rather than addressing fundamental glass processing hazards, and significant time was lost educating the consultant on process hazards of fiber glass production.

Since the initial use of PHA in a glass operation, the process has evolved and JM teams have identified and deployed a variety of different tools including "What If" analysis, Failure Mode and Effects Analysis (FMEA), and the Hazard and Operability Study (HazOp). JM's experience shows that an optimal PHA approach is a combined function of the nature of the process under evaluation, the maturity of the design, and available process safety information.

What-If Analysis

The "What-If" analysis is a straightforward process in which a proposed design is broken down to its core components and specific failure scenarios of these components are reviewed (e.g. what if cooling water flow is lost to a specific system component?). The analysis team typically includes a broad cross section of stakeholders including designers, scientists, engineers, production managers, production employees, maintenance managers, maintenance employees, construction team members, and health and safety professionals.

<table>
<tr><td colspan="11" align="center">Johns Manville
Example Manufacturing Facility 'What-If' Analysis
PHA Table</td></tr>
<tr><td colspan="11">Process section: Batching
PIDs: 010, 025, 100</td></tr>
<tr><td>No.</td><td>What if...</td><td>Hazard</td><td>Consequences</td><td>Safeguards</td><td>Severity</td><td>Likelihood</td><td>Risk Ranking</td><td>Recommendations</td></tr>
<tr><td>1</td><td>The high level switch LSH 0013 on the Internal Cullet Silo fails.</td><td>Cullet overflows the silo</td><td>Back up in the charge system</td><td>Plug detector LS-0011 in the bucket discharge

Zero speed switch SSL-0016 on the Internal Cullet Bucket Elevator</td><td>4</td><td>3</td><td>Low</td><td>131. Consider including the testing of the high level switch LSH-0013 on the Internal Cullet Silo in the mechanical integrity program.</td></tr>
</table>

Figure 2: Example "What-If" Analysis Table

The "What-If" analysis is led by a facilitator who specifies the process scope, defines quantifiable hazard characteristic categories (such as severity of incident or likelihood of occurrence), and facilitates an analysis of identified failure modes. In a glass plant, typical questions might be, "what happens if a forehearth develops a leak at a specific feature?" or "what if a furnace level detector fails?" The team then identifies the hazard associated with a given failure and the consequences of such a failure, identifies existing safeguards, classifies the potential severity of the hazard, determines the likelihood of the failure occurring, and defines the resulting risk ranking. Mitigation recommendations are then developed as merited by the risk ranking. JM has learned that "What-If" analysis is most effective when teams have intimate knowledge of the processes under evaluation and experience operating such processes. "What If" analysis is the most effective approach to physical projects when the PHA must be conducted early in the design process (e.g. completion of first drawings).

Failure Mode and Effects Analysis

The Failure Mode and Effects Analysis (FMEA) process is similar in many aspects to "What-If" analysis in that identification of failure modes is unconstrained by the tool format. The FMEA approach utilizes a team to identify critical process steps and then seeks to explore how those steps can go wrong (the failure mode). For each failure mode, the team documents the effects of failure as well as the potential cause(s) of the failure. Any given process step may have multiple failure modes. FMEA differs from "What-If" in that the "What-if" analysis inherently defines the mode of failure and carries the analysis from that point. The FMEA tool provides a structure for risk ranking (the Risk Prioritization Number or RPN) as the product of numerical scoring of the severity, potential for occurrence and probability of detection.

In Johns Manville's experience, FMEA works exceptionally well with projects in which procedural changes are being evaluated. For example, FMEA can analyze potential failure modes from changing a process sequence to mix a given fluid or from changing the way in which an operator performs a given check of a gauge. FMEA and "What-If" analysis both have an aspect that can serve as either strength or weakness. Both processes rely heavily on existing knowledge of the process under review and the ability of team participants to provide that knowledge during a review. As team participants may vary in their knowledge of process operation and potential failure modes during any given analysis, successful use of both FMEA and "What-If" analysis require an experienced facilitator who is both knowledgeable of the process under review and capable of extracting all available knowledge from participants.

Hazard and Operability Study

A Hazard and Operability Study (HazOp) approaches process hazard analysis slightly differently than the "What-If" and FMEA techniques. HazOp addresses the analysis by breaking down the proposed process to a specific operational node (e.g. glass furnace, natural gas system, burner #1) and addresses a list of specific operating parameters (e.g. flow, pressure, temperature, level) for possible deviations (e.g. too high, too low, stopped, reversed). For each combination of a parameter and a possible deviation, a cause (or causes) is identified as well as the consequence(s) of such a condition or conditions. Safeguards and recommendations are then specified and recorded and a severity and likelihood assigned. Relative risk is determined as the product of severity and likelihood.

The HazOp approach has proven quite reliable in risk analysis for manufacturing equipment; however, HazOp is not useful in the analysis of changes such as procedural changes or personnel changes. One of the challenges of HazOp analysis is that process safety information such as line drawings, materials of construction, procedures and similar information need to be mature in their development in order to apply the technique effectively. This drives the timing of application of HazOp to a point later in the project development cycle and can throw significant projects off schedule should unacceptable situations be identified. When applying HazOp analysis to significant projects it is therefore imperative that timing be built into the schedule late in the development phase for potential design refinements. The HazOp approach is beneficial in allowing analysis team participants with lesser levels of failure mode knowledge and experience to be productive participants. Team members are prompted to consider a variety of potential failure modes, assess whether they could happen, how or why the failures could occur, and develop recommendations to lower the associated risk. Unlike

"What-If" or FMEA, success of the HazOp process is less reliant on a facilitator guiding the analysis team to identify potential failure modes.

Location/Unit: Node: P&ID: PFD: Date: Session:	Example HazOp Analysis 1, Heat Exchange and Transfer 05195-PI 11/21/2009 1		Equipment #: Description:	Hot oil heat exchange and transfer process Oil system between main pumping skid and transfer to process							
Parameter	Deviation	Cause	Consequence	Safeguard	Recommendation	#	S	L	RR	Due Date	Assignment
Flow	too high	See Hot Oil Main Skid									
Flow	too low	Isolation valves in wrong position	Oil not heated, lose heat transfer at cans, poor product, scrap	Oil temperature continuously monitored and alarmed	No further recommendation	4	3		7		

Figure 3: Example HazOp Analysis Table

PHA Outcomes

The Johns Manville PHA process outlined above has been applied numerous times to projects ranging from modification of a gas burner purge system to the design of a light density wool fiber glass production line to an analysis of an entire operating facility. JM experience with the PHA process indicates that the process delivers benefits in three areas: 1) safer systems are initially implemented, preventing injuries; 2) JM avoids revisions late in the design process, lowering costs; and, 3) optimized solutions are designed into new processes and systems.

Johns Manville uses the PHA process to identify very specific design issues likely to result in loss at some point during asset life. For instance, during one PHA, the team found an extensive natural gas system for a production unit lacked an effective emergency natural gas shutoff. In another instance, use of PHA identified a design flaw in a cooling system that allowed a "short circuit" around the actual cooling tower. This flaw represented a covert design failure mode that may have gone undiscovered during system checkout, potentially resulting in catastrophic equipment failure on startup. For Johns Manville, conducting PHA sessions successfully achieves the core intent of delivering safer systems.

In addition, by drawing on participants from across a wide spectrum of functions, the PHA process improves the ability to design and install systems with more intuitive operability and greater maintainability. Using PHA, design stage configurations that may have proven confusing to operators were identified and resolved prior to construction. Equipment designs that would have been difficult to maintain were corrected, and sub-optimized systems have been re-designed for improved safety prior to construction. Examples include a control set-up in which green lights indicated both open and closed valve positions, a drive transmission mounted in an awkward position resulting in ergonomic exposure, and a frequently accessed hazardous energy isolation point being mounted too far from the operator platform. By formalizing a process to identify and mitigate issues in the design stage, JM realizes significant cost savings by preventing later revisions to equipment, and more importantly, avoiding medical and other human costs associated with an injury.

Finally, the PHA process promotes creative and insightful process optimization opportunities that may otherwise go unrealized. Examples include opportunities to eliminate complex logic and valve configurations with insightful weir and channel arrangements, and the realization of other improved design alternatives following analysis of traditional designs that involved high maintenance requirements. PHA provides a forum through which engineers, operators, and maintainers can participate in objective-driven discussion and arrive at optimized solutions to the real-world challenges presented by complex manufacturing systems.

MANAGEMENT OF CHANGE

The preceding section described how Johns Manville uses a PHA process to ensure that new operating equipment is designed and installed with the best organizational knowledge available. PHA is effective for new equipment, but a majority of safety incidents involve older assets and systems. JM still required a tool or process to allow review of existing assets with similar diligence to the PHA process. Again, JM looked to other industries for best practices and found that the extensive and successful use of the Management of Change (MOC) process in the chemical industry would be an excellent way to evaluate and reduce risks from existing assets in JM glass operations and other manufacturing platforms. The JM adoption of MOC involved defining specific thresholds for changes that would trigger analysis, developing a tool to review specific hazards, and in order to ensure appropriate hazard mitigation steps were completed, establishing an integrated pre-startup review process.

Critical Change Thresholds

To develop a sustainable process, definitions were required for repeatable and reproducible thresholds of change, which would consistently trigger application of a review process. The applicability of the MOC process to changes also had to be carefully defined to support organizational progress without subjecting all changes to excessive analysis. A balance was required between the defined triggers and the nature of changes subject to MOC processes such that the process would provide tangible value to operations leaders (plant managers) and would function without continuous policing through some form of corporate oversight. An additional challenge was to develop the process in such a way that it could be universally applicable to the manufacture of glass products, roofing products, machining operations, polyvinyl chloride extrusion processes, polyester and other synthetic nonwoven fibers and other operations carried out by Johns Manville.

The JM development team for the MOC process began by defining "change" within Johns Manville processes. The final definition of change included physical activities involving processing facilities and equipment, software and logic changes that might alter automated decision making, procedural activities that might affect personnel, process steps or inspections, or policy and organizational changes that might result in environmental, health or safety impact. Design explorations were consciously excluded from the definition of change.

Next, the definition of change was split into activities defined as either critical change (which would trigger MOC analysis) or non-critical change. Critical change is defined as change involving critical activities, critical systems, or critical materials. Once these categories of change were developed, the team developed checklists to serve as comprehensive and exclusive inventories comprising each of these three elements. A proposed activity involving a critical element would be subject to the Management of Change process and proposed activities falling outside of the listing would not be required to go through the process.

Only the key elements of the MOC process are detailed above since an exhaustive discussion of the entire process is beyond the scope of this document. Other features that will be mentioned in passing here include the need to specify responsible, accountable, consulted and informed parties in order to ensure thorough stakeholder inclusion. The process must also consider emergency response (such as if a failure occurs on a holiday weekend), how the process accommodates routine replacement (an MOC cannot be required for routine maintenance situations), documentation features and similar considerations.

Critical Change Analysis

Once a proposed activity has been identified as triggering formal MOC review, a tool was needed that was more appropriate than PHA for the broad range of envisioned changes. A checklist was developed containing analysis elements or subject categories to guide users through the change analysis. Within each element an array of questions is framed for the analysis teams to consider in characterizing the potential hazards associated with a particular change. The analysis teams build the comprehensive hazard understanding from the individual consideration questions. In this way, a hazard that may otherwise be overlooked gets recognized. For example in glass production, a change in burner systems may be intended to lower cost, but if the additional Btu available is instead used to increase glass throughput, environmental considerations may need to be addressed.

For each individual question, the MOC process requires documentation in the form of an answer to the question, and identification of the assignee responsible for handling action items. The process assigns predefined action items based on the answers to the questions. This allows for a variety of mitigation responses appropriate to the level of hazard/issue identified. Responses to certain specific questions automatically trigger a requirement to perform a PHA. An example from the JM MOC process is that a PHA is automatically required to assess hazards associated with a change that has the capacity to negatively alter process temperature.

Pre-Startup Safety Review

Johns Manville requires completion of the MOC analysis prior to the authorization of funding on certain projects. By inserting the MOC analysis at the start of a project, JM obtains a higher level of assurance that an appropriate level of safety consideration is present from the start of a project. Additionally, structures are established to prevent gaps from developing between MOC-generated actions designed to reduce risk, and what the project actually implements. One such structure is the Pre-Startup Safety Review (PSSR) process, which ensures that issues identified in the project funding stage MOC analysis are executed prior to commencement of operation.

JM developed its PSSR process to evaluate a list of discrete considerations that JM considers essential to the controlled commencement of operation of a given project or change. These items were broken into five sections: Environmental, Health and Safety (EHS); Asset Care and Reliability (AC&R); Procedures and Documentation; Training and Information Systems. Each item is analyzed for applicability, requirements for start-up, completion and comment.

The Pre-Startup Safety Review checklist requires an assessment of whether an item is required prior to startup, and also whether the item is completed. This has been found to be not only appropriate but necessary to ensure the tool remains effective in a pragmatic way. Consider, for instance, the installation of a mezzanine with three access stairways. The stairs are pre-fabricated with installed handrails and delivered to the facility with a gray powder coat finish. In an effort to promote actual use of handrails, the factory considers it essential to paint all handrails in a "safety yellow" color. Is it necessary to have all three sets of handrails painted safety yellow prior to startup? Maybe not – if the project is a multi-million dollar effort, it may be cumbersome to stall commencement of operation for every "nice to have" feature. This defined PSSR approach has resonated across the JM organization to provide features that combine the spirit and intent of enabling exceptional safety outcomes with the pragmatic ability to provide for such outcomes using a number of approaches.

MOC Process Evolution

Successful application of the MOC process provided plant leaders with the ability to better position their teams for efficient and safe operations. After full implementation of the new MOC process worldwide in 2008, JM found that the entire MOC review process or specific elements of the process were being applied to projects that did not meet the definition of "critical changes" This happened simply because the process consistently enabled projects with superior outcomes. JM leaders analyzed feedback from the facilities from process audits and reviews during 2009 confirming that the MOC process was effective, but the process was still not as efficient as it could be. The time commitment required for each MOC review was too long.

In response to the feedback, in 2010 JM rebuilt and re-categorized the MOC checklist for better hazard analysis and better integrated checklists used for PSSR into the main MOC checklists. This eliminated some duplication of work. The revisions to these MOC tools have reduced redundancy and eliminated the need to invest time in areas of little relevance, resulting in a much more efficient process. The MOC process at JM continues to have broad support and provides the operations teams with a streamlined tool to arrive at enhanced operational safety and effectiveness.

MOC PROCEDURE REFERENCED SECTION:			SECTION 2							SECTION 3		SECTION 4
No.	Key Word	Checklist Item	Is item applicable? Yes / No	Design Phase	PSSR Phase	Required for start-up? Yes / No	Action Item	Responsible person for action item	PSSR Walk-Thru: Item complete? Yes / No		Comments	Completion Date
		ENVIRONMENTAL										
ENV1	Air Emissions	Require modifications to air emission or air emission abatement systems that alter loading, compositions or fuel sources.										

Figure 4: Example MOC Table

TECHNICAL INQUIRY

Historically, JM investigated process safety incidents, evaluated causes, and made corrections. However, the ability to identify and integrate learning from these events into sustainable risk reduction relied almost exclusively on the investigative approach, skills, and capabilities of the local factory team. Such an approach yielded a variety of results ranging from in-depth and insightful learning to marginal and superficial glimpses into causation.

In 2005, JM had a significant process safety event involving a glass leak from an electric glass melter. Although there were no injuries associated with the incident, the company revised its response and investigation process and approached the incident with several objectives. The first objective was to understand, in detail, the cause of the incident and develop and implement corrective actions which would help ensure that such an event would never be repeated. The second objective was to develop a formalized contingency response approach. This would enable JM to rely on a well-documented process for investigation in the event another process safety incident of similar circumstances or magnitude should ever occur.

On notification of the incident, an investigation team was formed consisting of a team leader and four team members representing a cross-section of design engineering, process engineering, operations and health and safety expertise. The local factory personnel supported this investigation team. The inquiry was broken into several elements. The first element involved employee interviews while the incident was fresh in the minds of involved workers. To this end, the team conducted joint interviews (five interviewers, one interviewee) of each employee involved in the incident response. The interviewing team asked employees very detailed questions in order to triangulate facts derived from all interviews. Each interviewer maintained their own detailed notes that were used to build a timeline of events that transpired very rapidly over the course of several minutes. The team resolved instances of conflicting versions of events to eventually arrive at an accurate order of events.

The team also completed a demanding and methodical physical deconstruction of the incident scene. At the beginning of the incident response, the team gave specific direction to factory personnel to preserve the scene of the incident. This included cordoning off the site of the incident, which was not disturbed prior to arrival of the team. Following completion of employee interviews, the team completed an FMEA identifying 60 discrete potential investigation failure modes and developed mitigation plans to address each potential failure mode. As examples, the team worked with a structural engineer to ensure safety of the damaged structure and with electricians in order to ensure electrical energy was appropriately isolated prior to commencing investigative work. Once safety of the incident investigation scene was assured, the team began the actual scene investigation. During the investigation process the team mapped the incident scene using photographs, hand drawn maps and elevation level sketches. As glass was removed, images and sketches of the findings were taken to enable later incident scene reconstruction, including object orientation in three dimensional space. Objects removed from the melter were photographed, tagged, assigned an "evidence identifier" and placed in a secure room. Through this process, the vast majority of parts involved in the operating melter were reclaimed, identified and retained for further forensic evaluation.

On completion of the scene investigation, the team began independently compiling identified facts. The team used Post-it notes to create an open-ended working timeline that had been drawn on a white board. Elements of the investigation considered conjecture were removed and placed on the whiteboard above the timeline. When all facts had been posted and arranged, the team had compiled 83 sequential facts. On a second timeline below this 83-point timeline of facts, the team compiled the employees' understanding and actions in response to the situation as it unfolded. The completed "situation map" told a very detailed story, but did not yet fill in all of the gaps relating to causation. In order to build a unified incident theory, the team took turns postulating causation and sequence of events in such a way as to formulate a single story that accounted for all known facts without introducing any elements inconsistent with those facts.

This approach did not lead to a single unified hypothesis of the occurrences that led to the failure but instead to four plausible scenarios that could have led to the facts and events as observed. With the team unable to discount any of the hypotheses, the team worked on actions that would prevent a reoccurrence regardless of which hypothesis was correct. The team created 63 individual action items that were then standardized on electric melters across all operating units in JM.

Since the first use of the Technical Inquiry process for the 2005 incident, Johns Manville has successfully standardized the technique for incident investigations. Emphasis is placed on a balanced investigation approach that equally weighs physical evidence and interview testimony to arrive at a factual timeline. Where appropriate, JM engages outside experts with particular expertise to assist the investigation team. Team members independently offer their incident hypothesis framework in the level of detail they find justified. Hypotheses are not debated as offered, but simply recorded. Each hypothesis is subsequently compared against the facts. The team then asks whether there are any facts unaccounted for or in direct contradiction with the hypothesis. If the proposed causation is consistent with the physical facts, the physical evidence is then measured against the hypothesis for inconsistency or direct conflict. Once all hypotheses have been discussed, and invalid hypotheses eliminated, remaining hypotheses are checked against proposed corrective actions to ensure corrections can effectively ensure they will prevent recurrence. Final approved corrective actions are then deployed across all similar technologies/situations to collectively advance the safety process. The Technical Inquiry is documented in a detailed report that is prepared using a common format and level of detail. The Chief Executive Officer of JM personally reviews all Technical Inquiry reports and approves the required action items.

SUMMARY

Johns Manville evolved three very distinct processes to ensure that hazards associated with its installed asset base are corrected as the technology evolves. These processes help assure that new assets are designed and installed with the objective of ensuring that the company's newest technology is also its safest and most efficient, and they help JM as an organization to learn from significant process safety incidents in such a way that recurrence is prevented. The Johns Manville tools to achieve these outcomes include use of a formalized Management of Change program, Process Hazard Analysis process and a Technical Inquiry approach that is repeatable and reproducible. In conjunction with employee engagement processes and regulatory compliance efforts, the organization's balanced safety efforts have delivered a sustainable step change improvement in workplace safety to help ensure its work force returns home each day in the same state of health as when they arrived at work.

Author's Note
This work documents the processes and efforts brought to life by JM employees across the globe all of whom need to be recognized for their commitment to safe and sustainable work practices. The author would like to thank Nick Brueckner, Mark Fidishun, Monika Nagl, Jawed Asrar, Mark Charbonneau, Melody Dunbar, Barb Menard, Scott Pusey, Jim Smith, Tim Swales and Phil Tucker for their support and assistance in preparing this work for publication.

End Note
PROCESS SAFETY MANAGMENT
In December 1984, an industrial incident in Bhopal India resulted in an immediate death toll of 3,800 (estimate provided by the state government of Madhya Pradesh as published on Union Carbide's Web site). Estimates of the total deaths and disabling injuries caused by the incident vary; however, figures of 15,000 deaths and 4,000 permanent disabling injuries are prevalent (see Worst Industrial Disaster Still Haunts India, www.msnbc.com) and approximately 500,000 injuries of lesser severity due to the release of a gas cloud containing methyl isocyanate. In response to this disaster, U.S. lawmakers sought regulatory provisions intended to prevent similar incidents from occurring in the United States. In 1992, OSHA promulgated the Process Safety Management of Highly Hazardous Chemicals regulation (29 CFR 1910.119), commonly referred to as the PSM regulation. The focus of this regulation is to prevent and/or minimize adverse outcomes associated with catastrophic releases of toxic, reactive, flammable and/or explosive chemicals.

Most of the elements of the PSM regulation are largely invisible to the glass industry because glass production does not typically utilize any of the PSM-regulated chemicals in appreciable enough quantities to trigger applicability of the PSM regulation. The PSM elements do have great value in risk management and safety professionals will clearly recognize that some of the PSM elements were extracted and successfully implemented to improve Johns Manville safety programs and effectively reduce risk in glass plants.

Refractories and Recycling

EXTRA CLEAR GLASS REFRACTORY SELECTION: A FOLLOW UP

L. Massard and M. Gaubil
Saint-Gobain CREE, Cavaillon, France

J. Poiret
Saint-Gobain SEFPro, Le Pontet, France

ABSTRACT

Extra clear glasses are widely used for solar applications. The extra clear glass properties (higher transmission coefficient, lower glass redox) would induce an evolution of glass furnace running conditions. As a consequence, the refractory interface temperature and glass flow rate will increase and the glass capability of dissolving oxygen bubble will be lower. Theses new conditions would affect the lifetime of glass furnace and glass quality.

New fused cast AZS has been developed for these applications to improve and secure the lifetime of glass furnace and the glass quality regarding oxygen bubbles and AZS stones defects. In this paper, SEFPRO will present the properties of this new product.

1. INTRODUCTION

Extra white glasses are widely used in industry as substrate or support for many solar applications such as photovoltaic panel, thermal solar or solar concentration.

The extra clear glass is characterized by lower iron content (<200 ppm) and redox values. In terms of property, theses characteristics lead to higher transmission coefficient due to higher glass thermal conductivity, as seen on figure 1.

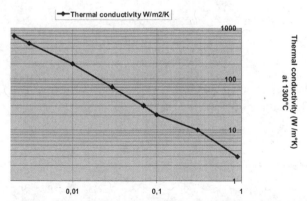

Figure 1: Thermal conductivity evolution versus FeO content at 1300°C

The higher transmission coefficient of extra white glass will modify glass furnace running conditions. At first, due to the rise of heat transfer into the glass, the refractory interface temperature should increase especially for the bottom soldier block and the pavement compared to usual conditions. Then, the thermal change should lead to an increase of glass flow rate (i.e. glass thermal convection) in EWG. Finally, due to low iron content, theses new properties could induce some changes on blistering due to the evolution on the glass/refractory interface and the low capability of this glass to dissolve oxygen bubbles.

In a point of view of glass furnace, it could lead to an increase of localized corrosion by upward drilling and decrease of glass quality.

2. CONSEQUENCES ON REFRACTORY

At the refractory point of view, these conditions may imply corrosion highly enhanced. Figure 2 presents pictures of used soldier block in extra white glass. We notice that the corrosion is characterized by high corrosion level of bottom block and an accelerated corrosion at block 'joints.

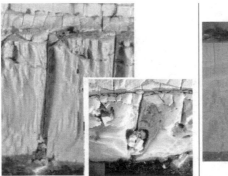

Figure 2: Examples of corroded soldier blocks

2.1 Corrosion of bottom block

The accelerated corrosion of bottom block could be explained by the increase of the corrosion speed linked to the rise of the refractory temperature interface and glass convection. From dynamic corrosion lab-test results, we can see, on figures 3 & 4, that temperature remains the main factor of corrosion with a exponential dependence of corrosion speed to the temperature and corrosion speed is proportional to square root of glass flow. The increase of corrosion could impact the glass quality by increasing stones and cords defects.

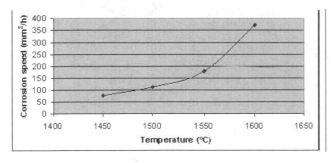

Figure 3: Corrosion speed vs. Temperature for AZS refractory (MGR results – 6 round/min)

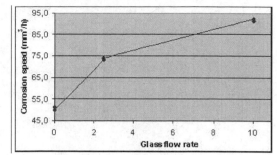

Figure 4: Corrosion speed vs. Glass flow rate for AZS refractory (MGR results)

Theses results show the high interest of increasing bottom block resistance by using "void free" soldier block.

In fact, as seen on figures below, the "void free" soldier block is characterized by good homogeneity in chemistry and density and induce, as a consequence, better corrosion resistance homogeneity. This fact has been validated by corrosion test (PFT).

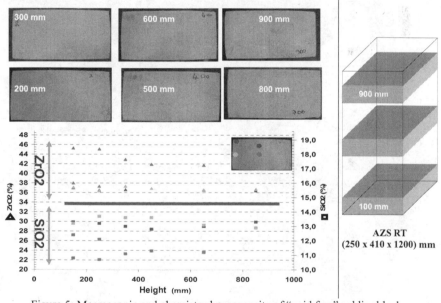

Figure 5: Macroscopic and chemistry homogeneity of "void free" soldier block

2.1 Corrosion of block' joint

Theses conditions may imply high localized corrosion that could be a source of unexpected glass leakage. The corrosion is characterized by an accelerated corrosion at block joint by upward drilling process.

The "upward drilling" phenomenon inside the joint is mainly due to the presence of bubbles (raw materials, oxygen blistering sensitivity with EWG) and a "opened joint" with AZS material.

Concerning the oxygen blistering sensitivity, tests (crucible tests during 30 hours) in atmospheric conditions show the higher capability on AZS material in EWG (in blue) compared to standard glass (in red) especially at low temperature. Also, theses tests shows the negative effect on blistering of zirconia content in the refractory and low iron content of the glass. This phenomenon is emphasized for product with high zirconia content.

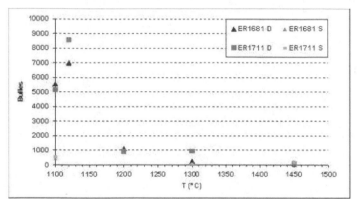

Figure 6: Blistering measurements for AZS ER1681 (32% ZRO2) and ER1711 (41% ZrO2) in EWG (in blue) and standard glass (in green)

The influence of atmosphere has been evaluated too. A blistering test with a Argon atmosphere has been realized and compared to air atmosphere. The test reveals that, by suppression of external oxygen, we can stop the blistering at low temperature.

So, the various results indicate that the "low temperature" oxygen blistering seems to be linked to an electrochemical process controlled by oxidation/reduction reactions, electronic conductivity of zirconia and external air.

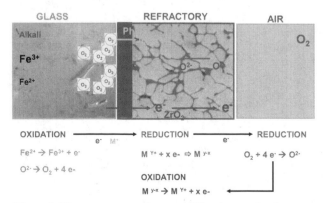

Figure 7: "Low temperature" oxygen blistering mechanism

This kind of phenomenon can occur in « zirconia based refractory » at block's joint. Due to thermal gradient inside the block and volume contraction imposed by zirconia transformation, we can observe some opened joint at high temperature. This phenomenon should be emphasized in extra white glass at the block's bottom due to the temperature rising. As a consequence, the glass can penetrate the joint and reaches some low-temperature area of refractory where zirconia is monoclinic so electronic conductor. So, the electrochemical blistering proceeds.

The consequences for glass furnace are the possibility to have more bubbles defects in the glass and to enhance corrosion by upward drilling.

3. NEW AZS REFRACTORY SOLUTION FOR EXTRA WHITE GLASS

To answer to these new conditions and complementary to the new self leveling refractory mortar, presented last year, new fused cast AZS, called "ER2010 RIC", has been developed for these applications to improve and secure the lifetime of glass furnace and the glass quality regarding oxygen bubbles and AZS stones defects. The table 1 gives the typical chemical composition of this product.

ZrO2	Al2O3	SiO2	Na2O	Others
36	Cplt.	14	<1.1	<4

Table 1: Typical chemical composition of ER2010 RIC

The product is dedicated to the closure of the block's joint, the decrease of blistering capability at low temperature, and the improvement of corrosion resistance. These objectives have been reached by the addition of yttria. Yttria contributes to dope zirconia grains and as a consequence to modify the thermal expansion curve. The new curve is characterized by a lower zirconia transformation temperature which allows to obtain the joint closure at running temperature and to reduce the domain where zirconia is an electronic conductor.

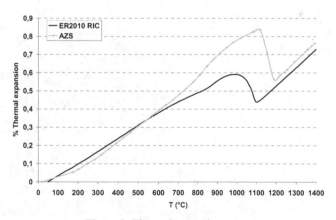

Figure 8: Thermal expansion curves

Regarding to the blistering (crucible tests during 30 hours), the figure 9 & 10 show that the capability of these new product is highly reduced, especially at low temperature.

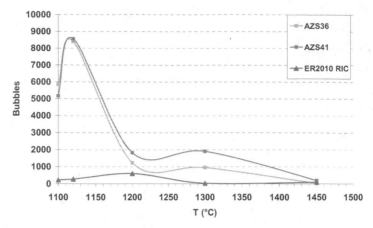

Figure 9: Blistering results in EWG (crucible tests during 30 hours)

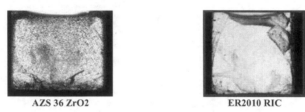

| AZS 36 ZrO2 | ER2010 RIC |

Figure 10: Comparison on blistering at low temperature

Finally, with the new chemical composition, the corrosion resistance has been improved compared to standard 32% AZS for bottom paving application.

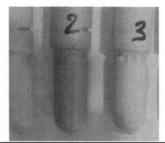

		AZS 32% ZrO2	ER2010 RIC
Total corrosion	V (cm3)	9.05	7.77
	Index	100	116
Flux-line corrosion	V	1.76	1.39
	Index	100	126

Table 2: MGR results at 1500°C during 48 hours in EWG

We can notice that this product has a good industrial feasibility and is actually in test as soldier block in extra white patterned glass furnace.

4.   CONCLUSION

Extra clear glasses are widely used for solar applications. The extra clear glass properties and especially the higher transmission coefficient would induce an evolution of glass furnace running conditions. As a consequence, bottom furnace refractory corrosion (bottom soldier block and pavement) may be enhanced. We have shown that the corrosion increase is mainly due to combined factors such as temperature and glass flow rate, "low temperature" blistering sensitivity and opened joints.

To solve this problem, new fused cast product, called "ER2010 RIC", has been developed for soldier block and bottom paving application with a better close join, low blistering capability, an improved corrosion resistance. This new product will contribute to improve and secure the lifetime of glass furnace and the glass quality regarding oxygen bubbles and AZS stones defects.

Finally, SEFPRO can propose optimized solutions for extra white glass application that consist of using:

- for soldier block, the new AZS solution associated with "void free" filling (cf. figure 11),

- for paving application, new fused cast AZS tiles with a optimized joint machining (TJ) in glass contact, and a new AZS mortar layer that allow to close the thermal expansion join after heating up and present a better corrosion resistance (cf. figure 12 & 13).

Figure 11: Pictures of ER2010 RIC "void free" soldier block

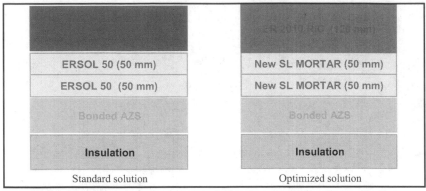

Figure 12: New optimized pavement solution for extra white glass application

Figure 13: Pictures of the SEFPRO' optimized solution for paving application

REFRACTORY ISSUES AND GLASS PROCESSING AND PREVENTATIVE SOLUTIONS

Paul Myers
Principal Consultant- Refractory materials
CERAM

ABSTRACT

This paper will review some typical refractory-related problems within the glass sector. The author will examine symptoms, investigate root causes and make recommendations on corrective actions. Furthermore, advice will be offered on how best to minimise and avoid problems.

Through a careful programme of testing of refractories for a given installation in-service issues can be minimised. In addition, the acquisition of such test data can assist in corrective measures should problems ever arise, thereby speeding up the problem-solving process. Using the skills and expertise of a reliable consultant engineer prior to the commissioning of a job can alleviate process and operational issues. Using modelling techniques such as FEA (*Finite Element Analysis*) to simulate situations before and after commissioning can identify and quantify the value of potential design changes, improve efficiency and often avoid damage to refractory linings.

1. INTRODUCTION:

Often refractories costs are quite a considerable chunk of the capital required to build a furnace. Consequently, this is the area where people most frequently look for cost reductions. Sometimes this can lead to in-service issues that present a false economy, as people look to use cheap alternatives and unqualified suppliers. This, in turn, may lead to inferior performance, both of the refractory materials and of the furnace itself. The costs of repairs in-service can be very severe, especially when the expense of production downtime is taken into consideration.

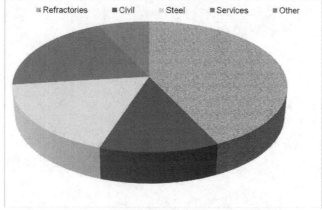

Figure 1: Breaking up of the building costs of a glass furnace.

If one is looking to make cost reductions in refractories it is better to do so through a careful programme of work where the key performance criteria of the materials are considered. The best way of doing this is through the measurement of the main properties that are specific to a given application and then the ranking of these measurements against known standards that have been run in previous campaigns. Additionally, some type of simulative testing can be undertaken that mirrors the main demand placed upon refractories in a given application. As a further insurance policy, technical audits can be undertaken to validate new suppliers and new materials. This higher level of consideration rules out variation within production campaigns for refractories and allows the glass maker to understand the real technical capabilities of any given supplier.

In following the procedure outlined above the glass maker can understand the true costs and implications of a material change before making an engineering decision. This approach will save the glass maker money in the long run.

Generally all refractory problems can be avoided if upfront investments are made to understand application and materials used. Good practice is for the glass maker to have a list of approved suppliers of materials that are bench marked with the key performance criteria and material properties. This generally will serve as a blueprint to qualify material changes. Strong partnerships are advisable with refractory suppliers; this allows all the main issues to be thoroughly understood. Quite often refractories are selected on limited knowledge of the material properties; this does not always give the refractory supplier the right definition of specification to fully service the glass maker's needs. In working closely with suppliers other potential side-effects of material changes can be ruled out also.

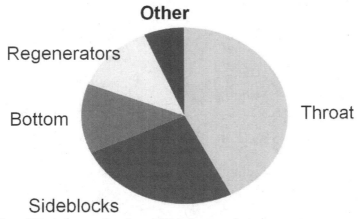

Figure 2 Typical areas of weakness of glass furnace refractories.

When testing materials ensure that truly representative production samples are used; ones that have been specially prepared by the manufacturer for a suite of tests should not be used. Make sure that samples are a true reflection of the manufacturing process by taking samples of actual products. Though this sometimes can be more costly in the short term, it does pay dividends in the longer term. It is important to make sure that a carefully planned sampling protocol is followed, so that the correct section of material is being tested. For example, with fusion cast products there is considerable variability within the microstructure that is dictated by the solidification process and is unavoidable.

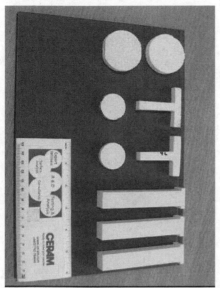

Figure 3 Sampling of fusion cast refractories; test coupons.

It is obviously important not only to test, but also to understand the results of those tests. If expertise does not exist in-house then experts should be consulted. It is important to spend time upfront to make sure that the testing criteria is clearly defined and understood. Where possible, well-documented material standards should also be incorporated into the programme as a control.

In the unfortunate event that a problem is encountered, then understanding the root cause of the issue is crucial. More often than not the glass maker will simply change supplier if they come across problems without really understanding the root cause. The financial pain of the problem may leave a very bitter taste at the time but it is important to really understand the causation and implement this into future selection criteria to make sure that lessons are learnt and do not let happen again. In refining selection criteria a future supplier can be guided to service refractory requirements.

2. REFRACTORY ISSUES IN GLASS MAKING:

2.1.   Contact materials

Fundamentally, the biggest issue is often glass corrosion of the refractory as this dictates the service life of the application. Secondary to this though is the impact of the refractory on glass quality. This is, of course, of great importance to the glass maker as glass quality issues can lead to costly production losses. Glass quality matters can often be overlooked by the refractory producer as it is something that is not clearly defined in the specification of the material, making it very difficult for them to control.

Glass corrosion can be determined and ranked through the application of tests such as the static finger test and dynamic finger corrosion tests. Of the two tests, the dynamic finger procedure provides data that is much more closely linked to real in-service performance of the refractory.

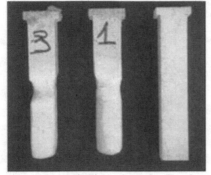

Figure 4 Dynamic Finger Corrosion Test Coupons.

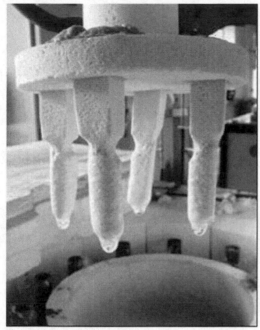

Figure 5 Dynamic Finger Corrosion Tests.

Glass quality issues can be screened out through various testing procedures which include techniques such as blister tests, bubble potential tests, stoning tests and fusion cast products exudation tests. This provides a further level of qualification for the materials, giving the glass maker confidence that they are not purchasing problematic materials.

Figure 6. Blister tests of refractory materials.

2.2. Material for structural applications

Fundamentally, we must ask the question - what role does the refractory have to serve in the application? What are the operating conditions that are placed upon it? Temperature, mechanical loads, thermal cycling, vapour exposure, gaseous environment, design of the product (could stresses be minimised by modifying design?) combustion system (and any related gaseous species) and demands/ variations within the glass making process? It is important to make sure that the selection criteria reflect the functionality and the key elements of the application.

Refractories in a roof will, for example, need resistance to creep and alkali attack. In an application such as a common wall in a regenerator the material will need to have a common resistance to creep, plus resistance to alkali attack in the upper zone of the regenerator and resistance to thermal cycling in the lower zone.

In structural applications it is very important to understand the warm-up behaviour of the refractories and incorporate this into the design. For example, thermal expansion of the material must be carefully engineered for in order to avoid either unnecessary gaps opening up that can lead to both the ingress of alkali vapour detrimental to the refractory lining and high stresses (if the expansion is not accommodated). Both of these could lead to failure of the lining.

Creep can be assessed in the laboratory through the creep in compression test where refractory specimens are subjected to load and temperature within a specialised rig. Thermal expansion can be assessed easily in the laboratory and is very important information for both furnace designer and installation engineers.

Alkali vapour attack tests can be performed on a laboratory scale and provide important information on how materials can withstand such an environment. Deterioration by alkali vapour can often lead to the spalling of refractories which can cause zones of roofs to fall in. Species such as nepheline (a feldspathoid; $Na_3KAl_4Si_4O$) may form within alumino-silicate type refractories; this has a high associated volume change and considerable thermal expansion mismatch which can lead to the detachment of the affected zone from the body of the refractory. Knowing and understanding the mechanical properties at the service temperature is very important; these can be determined using specially designed test rigs which incorporate furnaces that allow the measurements to be made at temperature.

There are various tests that may be applied to investigate the thermal shock resistance of a material or to assess its resistance to thermal cycling. Such tests include hot to cold cycling, hot to hot cycling and water quenching. In the hot to cold cycling test samples are placed within a specially designed furnace which has a hot zone that can be set to a given temperature and a cold zone which blows air on to the samples. The samples are then cycled between these two zones over a fixed period of time. Materials are then assessed both visually for the number of cycles it takes to fail and by using an ultrasonic measurement before and after a number of cycles. The drop in the ultrasonic signal is then used to investigate the material's response to thermal cycling. The speed of sound in a solid is governed by an equation that incorporates both the density and Young's modulus of elasticity of that solid. We know that the density does not change in most materials through thermal cycling, therefore only the Young's modulus has changed. The Young's modulus normally has a linear relationship with strength so this can used as a measure of loss in strength.

In the hot to hot variation of this test the cold zone is another furnace zone which is set to a cooler temperature; a similar procedure is then followed as with the hot to cold test. The specimens are then assessed both visually and with ultrasonic measurement. This type of test is often very usual in the assessment of regenerator checker materials.

Another variation of the hot to cold test involves applying a flexural load to the specimen after quenching the material in compressed air. Materials are then cycled to failure.

Finally, for materials that are fairly resistant to thermal shock, there are two water quench methods that can be used to assess materials. In the first method materials are taken from a given temperature and quenched into cold water. The materials are then cycled to failure. The second variation of this test involves measuring the flexural strength through a three point bend test and then taking a similar set of materials and quenching them into cold water and reassessing the strength after the quench. The retained percentage strength is then reported in this test and is used to assess the materials.

2.3.   Functional glass making refractories

These are refractory components, which are interacting with the glass making process and are products such as refractory expendables - orifice rings, plungers, stirrers and feeder tubes. In such applications of refractories both glass contact issues and some of the structural issues need to be considered.

Sometimes these materials can be flawed during the manufacturing process which can limit their lifetimes. On occasions this can lead to in-service failure and can be costly to the glass maker. In light of this fact, it is advisable to apply some means of non-destructive testing such as an ultrasonic test or the application of a light paraffin oil (or similar chemical), which drains into a crack that may not be visible to the naked eye.

3. REDUCING THE RISKS

There are really four key strategies for reducing the risks associated with refractories for glass applications. These are refractory testing, technical audits, independent inspection and computer modelling.

3.1. Refractory testing

Applying some of the tests discussed earlier will help to minimise the likelihood of refractory issues. Furthermore, they will screen out problematic materials prior to the commissioning of projects. Knowing the material's performance upfront is an important step in the alleviation of refractory issues. The weakness though in this approach alone is the statistical significance of the sampling. It is costly in time and money to test large sample populations with a whole suite of tests. One approach to ensuring that laboratory testing is robust and representative is through technical audits.

3.2. Technical Audits

This approach is based upon an in-depth study of the refractory supplier accompanied with some supporting testing. Through an audit the true capabilities of the supplier will be reflected. Within this exercise the consistency of products is explored in addition to the state, capabilities and throughputs of equipment, the management of the plant and the SHE (safety, health and environmental) policies of the operation. All of these are key elements to ensure consistent and on time delivery of quality product. Large cost savings in the long run can also be realised by the glass maker.

3.3. Independent QA of installation and refractory sourcing

Such a project involves either (or both) the installation of refractories and the manufacture of refractories being overseen by an independent party. This presents an unbiased view as to best practice which can then be followed by the operators. Costly mistakes could be flagged up immediately and taken out of the process. Furthermore, when someone is been observed it is always human nature to do the best possible job!

Often, when using a new supplier, this type of approach can iron out any potential teething problems of the new relationship that can be costly to the glass maker. In the exercise clear and measurable in process controls can be established, which regulate the supply of consistent quality refractories to the glass maker. Any factors that can impinge upon refractory lifetime can be highlighted and addressed.

Recent experience of such projects, in which CERAM has been involved, has led to glass makers realising cost savings of millions of dollars.

3.4.   Computer modelling and design verification

These techniques are incredibly useful for the furnace engineer when performing design modifications or when increasing the size of a refractory lining of an application. This allows the majority of unforeseen eventualities to be explored first, before the refractories go into service.

Generally the understanding of behaviour of refractories is limited within some companies and it can save money to use expertise when required.

Considerable modelling work is done within the glass industry for studying the flow of glass. However, techniques such as finite Element Analysis (FEA) can be used to investigate the flow of heat and development of thermal stress within refractory materials. This allows the in-service behaviour of refractories to be modelled; their designs can then be modified accordingly to minimise thermal stress, which means that the demand placed upon the material is lessened. This can lead to extensions to the lifetime of certain areas.

In FEA modelling, a thorough material model must be developed first - this includes measurements of Young's modulus of elasticity, Poisson's ratio, Failure stress (both compressive and tensile), coefficient of thermal expansion, thermal conductivity, density and specific heat capacity. These properties need to be determined with respect to temperature. Sometimes the material model is overlooked and this is a major pitfall of this technique. CERAM has considerable experience in the testing for such properties and in the development of material models. A CAD drawing of the refractory can be imported into the finite element software package and the material model applied to this. Either the individual component or assembly of refractories can then be modelled.

4. CONCLUSIONS

Refractory issues are incredibly costly to the glass maker, though through sensible strategies the major of refractory issues can be overcome.

The glass maker needs to fully understand the economics of refractory purchasing - it simply is not just a case of selling price of the materials, the performance and financial of this relating to a given material must be understood.

Understanding the root causes of refractory related products and learning the lessons from issues, no matter how costly they are, is crucial.

When doing design changes and expanding the capacity of a furnace, the use of modelling techniques should be considered and advice sought from refractory experts.

REFERENCES

[1.] Refractories – A Global strategic business report – June 2007, Global industry analysts Inc.

[2.] Burning issues, Asian glass, pp. 55-57, Apr-May 2007

[3.] Longshaw N, Refractories for a global glass market, proceedings of the 69th conference on glass problems, pp. 89-100 2009

[4.] Longshaw N, Structure, microstructure and refractory performance, proceedings of the 69th conference on glass problems, pp. 249-261 2008

NOTE:

Guidance white papers covering the sampling and testing of refractories for glass production are located on the CERAM website.
Also guidance papers are available in relation to computer modelling of refractories.
www.ceram.com/refractories/

FUEL SAVINGS WITH HIGH EMISSIVITY COATINGS

Tom Kleeb
North American Refractories Company
Pittsburgh, PA

Bill Fausey
Owens Corning
Granville, OH

ABSTRACT

High performance, high emissivity coatings have been applied to the crowns and superstructures of wool glass, composite glass, and soda-lime glass furnaces. These coatings have increased the radiant heating component of glass melting, resulting in fuel savings in excess of 7%. This paper discusses how increasing emissivity results in lower fuel usage. Issues such as glass contamination, alkali resistance, and coating life are also discussed. The results of extended trials of these materials at Owens Corning and Libbey are presented.

BACKGROUND

High emissivity coatings have been used in industrial applications, especially power generation and chemical production for over four decades. These products, which contain high emissivity oxide materials, were limited to applications having maximum temperatures of 1200°C. At higher temperatures, the emissive properties of these coatings decreased steadily to such a degree that they provided no meaningful benefit and were not cost effective.

The current space shuttle fleet developed by NASA in the 1970s incorporated a lightweight ceramic thermal protection barrier to protect the skin of the orbiter. This siliceous composition was chosen for its unique combination of low weight, high strength, and excellent thermal shock resistance. The maximum service temperature of the ceramic barrier, however, was not capable of withstanding the temperatures up to 1650°C that the bottom and leading edges of the wings, stabilizer, and nose cone saw on re-entry. The initial shuttle design included the use of a high emissivity coating to absorb and re-radiate the heat in these areas to the colder atmosphere of the Earth, essentially isolating the ceramic tile from the hot plasma generated on re-entry. Since no commercial coating available at that time was capable of maintaining a high emissivity at re-entry temperatures, NASA developed high emissivity coating technology based upon synthetic, non-oxide materials that was not only suitable for temperatures encountered in aerospace applications, but was also suitable for high temperature industrial processes such as steel and glass making.

In the 1990s, NASA began designing the next generation of orbiters, updating all designs and systems including the thermal protection system (Figure 1). While the deployment of this generation of orbiters has been canceled, the technologies, including the improved high emissivity coatings, have been made available to industry through a licensing program. Binder systems compatible with every quality of refractory were combined with NASA's latest high emissivity materials to produce a family of high performance, coatings that maintain emissivities near that of a theoretical blackbody at all glassmaking temperatures. These coatings, marketed under the brand EMISSHIELD®[1], were commercially introduced to the glass industry in 2006 by North American Refractories Company.

[1] EMISSHIELD® is a registered trademark for coatings manufactured by Emisshield, Incorporated, and is covered by U.S. Patent 6,921,431.

125

Figure 1 – The X-33 Orbiter

EMISSIVITY

High emissivity coatings are not insulators. They are not barriers to the conduction of thermal energy through furnace walls. Insulating refractories are generally placed behind dense refractories at the cold face of refractory linings. While this reduces heat loss from a furnace, the amount of heat stored in the refractory is increased and the refractory materials must withstand higher mean temperatures. Because furnace linings act as heat sinks, valuable process energy is absorbed by the refractories and lost by conduction to the cold face of the lining. Additional convective energy held by the furnace combustion gases is lost up the flue (Figure 2).

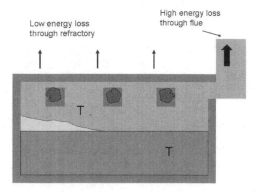

Figure 2 – Glass tank with insulated refractory superstructure and crown showing heat loss through the refractory and great heat loss up the flue.

When high emissivity coatings are applied to the hot face of the refractories in the superstructure and crown, radiant and convective energy from the burners and hot furnace gases are absorbed at the surface of the coating and re-radiated to the cooler glass batch (Figure 3).

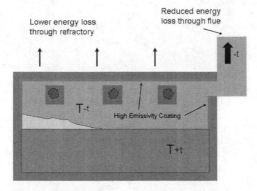

Figure 3 – Applying a high emissivity coating to the superstructure and crown results in lower heat loss through the refractories and significantly reduced heat loss up the flue. As a result, more energy is available to heat the glass.

For high emissivity coatings to be effective, the temperature of the coating surface must be greater than the temperature of the glass, which is always the case whether the glass batch is being melted or whether the molten glass is being refined. The amount of heat re-radiated from the coating is predicted by the following equation:

$$Q = E_w \cdot \sigma \cdot (T_C^4 - T_L^4)$$

Where: Q = re-radiated energy absorbed by the furnace load
E_w = emissivity of the coating
σ = Stefan-Boltzmann constant
T_C = coating temperature
T_L = load (glass) temperature

Since the temperature of the coating and the temperature of the glass are raised to the fourth power, it is apparent that the coatings absorb and re-radiate the most energy when the temperature difference between the coating and the load is the greatest. Therefore, the greatest opportunities for energy savings are when a cold furnace is being commissioned and in the area where batch is being melted during operation. The application of these materials above the melt line increases the radiative component of heating glass at the expense of the convective component. The coating absorbs convective heat from the hot gases and re-radiates this energy to the glass. The result is less energy being lost up the flue and more energy being used to heat the glass.

APPLICATION OF HIGH EMISSIVITY COATINGS

Unlike the use of insulating materials that have predictable performance characteristics under steady state conditions, the benefits of using high emissivity coatings depend greatly upon tank design and operating parameters. Uncoated refractories have emissivities, E_w, in the range of 0.4-0.6 at glass melting temperatures. The application of a high performance, high emissivity coating to the refractory increases the emissivity of the refractory to about 0.9 at glass-making temperatures. This means that about 90% of the energy absorbed by the coating is re-radiated to the cooler glass.

Referring to the equation on the previous page, it is easy to see that by increasing the E_w of the refractory, the heat absorbed by the glass, Q, will increase significantly. This is usually not desirable, where over-heating can change the viscosity and convection patterns in the molten glass and alter the entire production process, so something else in the equation must be reduced to compensate for the increase of E_w, to maintain a constant Q. The factor that must be reduced is the temperature of the coating and the furnace gases, and this is achieved by reducing the total energy input to the furnace. Of course, as total energy is reduced, fuel savings are gained. Theoretically, it is possible to consume the increased Q after a high emissivity coating is applied by increasing the pull rate. This would result in increased production at pre-coating fuel levels. While this practice is common in other industries such as ceramics manufacture and non-ferrous metals production, it has not been done in a glass furnace as of the writing of this paper.

After a high emissivity coating is applied (Figure 4), the combustion products of firing, carbon dioxide and water vapor, do not absorb all wavelengths of the continuous blackbody spectrum emitted from the coating. The complete spectrum, however, is absorbed by the glass. The result of this is that the temperature of the glass is hotter than the temperature of the combustion products in the furnace atmosphere. It is essential, therefore, that the melting process be controlled by the temperature and behavior of the glass batch rather than the temperature of the furnace atmosphere. While critical furnace parameters should always be monitored, the burners should be operated at a level that insures that the glass temperature at the throat and the batch line location at the glass surface are unchanged from pre-coating operation.

Figure 4 – Spraying a high emissivity coating on the crown of an insulation furnace.

A key parameter for determining appropriate burner settings for Owens Corning is the location of the point where the glass batch is 100% melted. If historic burner settings are used after the coating is applied, the glass will be hotter and the 100% melt point, or batch line, will be closer to the doghouse than normal. The burners must be turned down until the 100% melt surface returns to its pre-coating location. The key glass temperature control point is the entrance to the throat. Adjusting burner output so that the glass temperature equals the pre-coating temperature will insure that drawing and forming processes beyond the throat will not be affected by the use of the coating. In using a high emissivity coating in a glass furnace, it is desired that no changes in temperature or glass convection in the glass melt be made. The major furnace operating parameter changes should all be above the glass line, namely lower burner settings, lower furnace atmosphere temperature, and lower flue temperature.

After coating the breastwalls and crown of a glass tank, the furnace and crown behavior during heat-up will be different than that experienced before the application of the coating. Early in the heat-up process, the uncoated glass contact refractories and the solidified glass in the melter will be the low temperature load that will absorb the energy radiated from the coated superstructure and crown. At this point in the heat-up schedule, the difference in temperature between the coating and the hearth will be a maximum. Very little heat will be conducted through the coating and absorbed by the coated refractories, so the degree of crown rise typically experienced at this point of the heat-up in uncoated furnaces is not likely to occur. As the melter and its contents get hotter, the delta T between the coating and the furnace load decreases and the crown will start rising.

When batch is introduced into an empty furnace, the delta T will increase and the crown will drop, somewhat. The crown will begin rising again as the batch heats and eventually melts. A drop and rise cycle will again occur as glass batch is introduced. The crown will stabilize when production stabilizes and at that point it can be sealed. Newly-sprayed tanks with frozen glass in the hearth will also exhibit little heating of the crown early in the heat-up schedule. As the glass heats and finally melts, the delta T will decrease and the crown will rise. Another drop-and-rise cycle can be expected when the glass batch is introduced. Again, the crown should not be sealed until production stabilizes. Furnaces with silica crowns will show less movement with raw material introduction above the quartz inversion temperature.

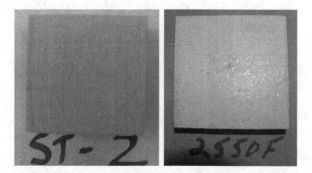

Figure 5 – On the left, a bonded AZS refractory plate coated with a high performance, high emissivity coating and dried. On the right, the same plate after firing at 1400°C.

ACCELERATED ALKALI TESTING AND GLASS DISSOLUTION STUDIES
Accelerated Alkali Test

The two most common concerns about applying high emissivity coatings to glass furnace crowns and superstructures is the alkali resistance of the coating and the issue of stoning potential and color change, should the coating fall into the un-melted batch or glass. To address the alkali corrosion question, a plate of VISTA® bonded AZS refractory was coated with the high performance, high emissivity coating most commonly used in glass furnaces. It and an uncoated plate of the same refractory were fired at 1400°C, then tested in an accelerated alkali test (Figure 5).

The test consisted of placing a quantity of sodium carbonate in alumina crucibles, covering the crucibles with the fired coated and uncoated bonded AZS refractory plate, then heating for 48 hours at 1430°C. After the heating, the cover plates were weighed, measured, cut, and examined (Figure 6).

 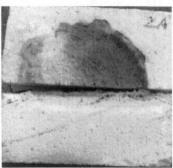

Figure 6 – Coated and uncoated bonded AZS refractory plates after the accelerated alkali test. The uncoated plate is on the left and the coated plate is on the right.

There was very little weight change between the uncoated bonded AZS refractory sample and the coated sample (Table I). The coated sample registered no dimensional change, while the uncoated sample showed a slight positive linear change. It was concluded that the high emissivity coating was at least as resistant to alkali attack as the uncoated refractory.

Table I – Measured Results of Accelerated Alkali Testing of
Coated and Uncoated Bonded AZS Refractory

Sample:	Bonded AZS Refractory, Uncoated	Bonded AZS Refractory, Coated
Weight Change, %:	+2.4	+2.3
Linear Change, %:	+0.4	0.0
Crucible Notes:	Major reactions and leaks	Minor reaction

Glass Dissolution Test

The glass dissolution test was a non-standard procedure where clear soda-lime cullet from an American container manufacturer was placed in an alumina crucible (Figure 7). A quantity of the high emissivity coating most commonly used in glass furnaces was dried and crushed. A piece of the coating was placed in the crucible, which was then heated for two hours at 1430°C. This was a static test where no attempt was made to homogenize the molten glass. After testing, the crucible was cut, the glass containing the coating was observed and samples were prepared for microscopic examination.

Figure 7 – On the left, the test crucible containing soda-lime cullet and dehydrated high emissivity coating before heating. On the right, the cut crucible after heating.

Examination of the fused cullet/coating sample revealed that the vicinity of the fused sample where the coating melted, but was not dispersed, had a light amber color. EDS analysis of this and other areas of the sample showed a greater silica content than the cullet where the coating was placed, but there was no evidence of stones or seeds anywhere in the sample. There was some reaction between the cullet and the crucible, as this interface had a greater alumina content than the cullet. It was concluded that the high emissivity coating does not present a danger of seed or stone formation. The actual amount of coating solids in a typical glass furnace is less than 50 kilograms, so this amount compared to the high glass production rate and the fact that the molten glass is homogenized in the furnace makes it unlikely that the coating would cause any discoloration of the glass, even if all of the coating was instantly lost.

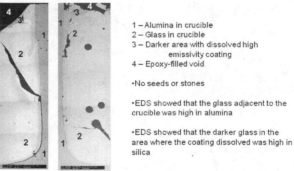

1 – Alumina in crucible
2 – Glass in crucible
3 – Darker area with dissolved high
        emissivity coating
4 – Epoxy-filled void

•No seeds or stones

•EDS showed that the glass adjacent to the crucible was high in alumina

•EDS showed that the darker glass in the area where the coating dissolved was high in silica

Figure 8 – EDS examination of the cullet/coating glass after cooling. No stones or seeds were apparent in the amber, coating-stained vicinity.

HIGH EMISSIVITY COATINGS ON OWENS CORNING INSULATION FURNACES

The first installation of a high emissivity coating on an Owens Corning insulation furnace took place in May of 2008. This installation coincided with the rebuild of a 150 metric ton per day oxy-fuel furnace in Owens Corning's Kansas City facility. The furnace was designated K-5. The previous campaign of K-5 was over a time period when Owens Corning was making use of progressively more purchased cullet in the batch. This allowed for a good data base to be collected over a wide range of cullet levels.

The top curve of Figure 9 shows the energy consumption in K-5 furnace at various cullet levels over the previous campaign. This plot simply confirms the well-established relationship that as cullet level in the batch is increased the energy consumption will decrease. These data points represent many months of operation at each level and thus are considered reliable for comparison to the energy consumption after installation of the high emissivity coating. The bottom curve is the energy consumption after the coating was applied to K-5 furnace. Cullet levels in the batch were changed to be able to compare directly to the pre-coating energy consumption. Care was taken to spend enough time at each cullet level to be sure any change in energy consumption was real. Clearly the energy consumption is less after application of the coating. The energy saving is in the range of 7-8%.

K5 % Cullet vs MM BTU/TON

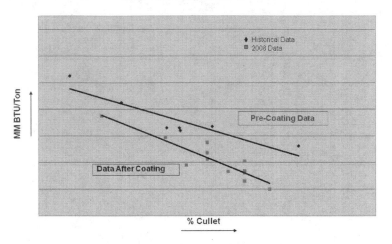

Figure 9 – Trend curve showing Gas Usage per Ton Versus Percent Cullet, before the coating was applied and after the coating was applied to the K5 tank. The axis values are proprietary and not shown, but the gas savings averaged 7.5% (range = 7-8%) over all cullet levels.

Of note in Figure 9 is the fact that the fitted lines are not parallel. The reason for this result is the lower purchased cullet levels were evaluated first. The furnace operators were aware of the potential for energy reduction but were cautious about removing energy from the melting system at first as removal of energy, from a historic operating view, would cause melting defects to appear in the forming operation. In short, "hotter" is safer! However, over the course of several weeks the operators became more comfortable in operating the furnace at lower energy input and a savings range of 7-8% was achieved.

It is fair to ask if K-5 had simply been operated at too high an energy input prior to the application of the coating. We can be sure this was not the case as specific changes in the front underglass temperatures took place after application of the coating. Specifically, the front underglass temperatures increased by 15 to 20 degrees Centigrade. Control of the furnace exit glass temperature is one of the parameters in this furnace control scheme and as front underglass temperatures were returned to their normal ranges the energy saving was achieved. Additionally, the position of the batch line was pulled back in K-5 furnace after start-up. This indicated to our operators we were transferring more energy to the batch and they responded by allowing the batch line to return to the pre-coating position in the furnace.

The data proved that energy savings were real after application of the coating. K-5 furnace was the first glass furnace to have the entire superstructure coated with a high emissivity coating. The second question then became; how long will the coating last?

K5 % Cullet vs MM BTU/TON

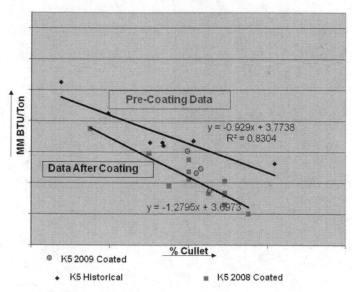

Figure 10 – Two years of gas usage after coating compared to historic fuel usage for uncoated furnace.

Figure 10 is a plot of the K-5 data for 2009 and 2010. Clearly the coating is still providing energy savings after two and one-half years of operation. Energy monitoring is completed monthly to follow the long term performance of the coating. Since the application and data collection on K-5 furnace, Owens Corning has applied a high emissivity coating to other oxy-fuel furnaces and to one gas fired furnace. Furnaces identical to K-5 have shown a similar 7-8% energy savings while other oxyfuel furnaces have shown savings as high as 10%. The application of a high emissivity coating to the superstructure of oxy-fuel and gas fired furnaces is now standard practice at Owens Corning.

HIGH EMISSIVITY COATINGS IN SODA-LIME FURNACES
Case 1

The first application of a high emissivity coating to a soda-lime furnace was in a Libbey gas-fired side port furnace melting clear glass. The application of the coating was made during a minor rebuild where all sidewalls, throat, port, and top checker repairs were made. One inch of additional insulation was installed on the bridgewall, and the flow meter was re-calibrated. The coating was applied to the silica crown only, which required no maintenance.

During heat-up, it was noted that more uniform crown versus bottom temperatures were achieved. Following the conversion to main burners, the fill began approximately 12 hours earlier and at a colder temperature than a traditional sequence. The fill rate was significantly higher than normal and operating glass level was achieved about 12 hours earlier than usual.

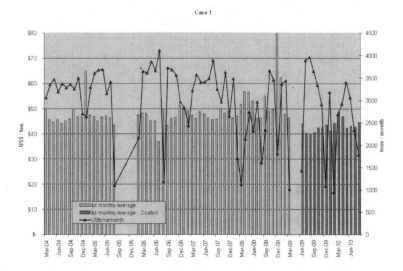

Figure 11 – Production and cost data for Libbey's Case 1 furnace before and after coating.

Figure 11 shows the production and fuel usage cost of this furnace in constant dollars for the first fifteen months of operation with the coated crown. Pre-coating data are shown for the previous six years. Boosting is used on this furnace intermittently, so production data for periods where boosting was used have been removed. The cullet ratio for the entire data set ranged from 35 to 45%. Evaluating a high emissivity coating which has been applied during a rebuild or maintenance outage is difficult. The evaluation of the coating is confounded by efficiencies gained by the rebuild. Libbey calculated that the gas savings due to the combination of the high emissivity coating and a traditional minor rebuild in the first 15 months of service was approximately 9%.

Case 2

The second Libbey furnace to employ a high emissivity coating was a dual recuperative, gas-fired furnace producing clear soda-lime glass. The coating applied was an experimental, non-oxide, high performance coating similar to the one used in their first furnace. This was also installed during a minor rebuild in the spring of 2010, which consisted of sidewall, throat, and recuperator repairs. The coating was sprayed on the silica crown only, which had seen prior service.

The heat-up and fill was done by an outside contractor using a standard protocol; no significant information regarding the heat-up is available. Figure 12 shows production and gas usage for the 2-1/2 years prior to the application of the coating and the preliminary fuel usage data for the first four months of operation with the coated crown. Approximately 6% in fuel efficiency was gained from the rebuild and use of the coating. Presumably some of this is due to the recuperator repairs, but Libbey has judged the use of the coating to be effective. A third gas-fired furnace melting clear soda-lime glass was sprayed with the same coating used in Case 1 by Libbey in June, 2010. It has not been in service long enough to yield statistically significant performance data, but early anecdotal reports suggest that its performance is mirroring that of the Case 1 furnace.

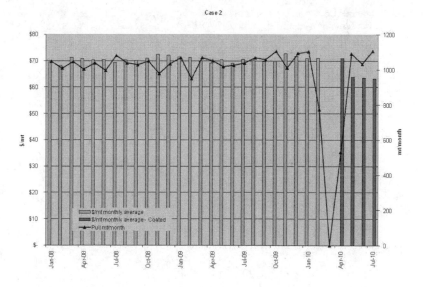

Figure 12 - Production and cost data for Libbey's Case 2 furnace before and after coating.

SUMMARY

In the past two years, seven furnaces melting wool glass, three furnaces melting composite glass, nine container and tableware furnaces, one lighting glass furnace, and two sodium silicate furnaces have been coated. The coatings used to coat these furnaces are actually a family of coatings that utilize the same emissivity technology. The members of this emissivity coating family each have binder systems developed for spraying on different refractory substrates. Refractory compositions that have been sprayed include alumina, mullite, all alumina-chrome compositions, silica, zircon, and bonded AZS. Because of the exudation of fused AZS refractories, they cannot be coated. Fused alumina, however, has been successfully coated in the laboratory. The refractory substrates to be coated do not have to be new, as long as they are not friable and are not coated with glass or batch materials. Because refractories containing chromium and zirconium have a higher emissivity before coating, it is likely that the fuel savings after applying a high emissivity coating to these refractories will not be as great as that seen when silica, alumina, and mullite refractories are coated. It is also likely that furnaces having the crown and superstructure coated will show greater fuel savings than furnaces with fused AZS superstructures where only the crowns can be coated. These suppositions will be confirmed or disproven as long-term performance data become available for the furnaces coated in 2010.

Users of high performance, high emissivity coatings can expect to realize lower fuel usage, which will lead to lower fuel costs, lower NO_X emissions, and a smaller carbon footprint. In the event that a glassmaker prefers to use the increased radiative heating provided by the coating to heat the glass faster, increased production at historic fuel levels are theoretically possible, provided that downstream equipment can handle the increased production rate. Additional benefits, such as longer refractory life and faster heat-up of cold furnaces have been reported.

ACKNOWLEDGMENT

The authors wish to thank Dan Cetnar and Libbey for their contributions to this paper.

REGENERATOR TEMPERATURE MODELING FOR PROPER REFRACTORY SELECTION

Elias Carrillo
RHI-Refmex

Mathew Wheeler
RHI-Monofrax

ABSTRACT
The current practice in the glass industry is to operate furnaces hotter, longer, and with higher pull rates. Glassmakers can no longer solely base their refractory selection and decisions on established and historical practices and guarantee previous refractory life. In this paper RHI presents modeling techniques that couple our regenerator efficiency program with standard modeling software. This method allows for tailor made refractory solutions; whether to pinpoint condensate zone locations or to establish proper material locations for temperature resistance enabling the ability to maximize performance, service life, and economy.

INTRODUCTION

The use of regenerators for improving the thermal efficiency of the glass melting process is a well-established method that dating back to 1867 with the Siemens invention, which at that time was a breakthrough in the steel industry. The regenerative furnace concept was then adopted by the glass industry and since that time it has had such robustness that it is currently the workhorse for glass furnace design all over the world. Despite the effectiveness of regenerators in the glass industry, the mathematical modeling of the heat recovery for characterization of a regeneration process [1] involves such a formidable task that actual regenerator calculations are undertaken and understood by only a few skilled designers using fancy software packages. Most of the glass producers rely on empirical rules and experience for regenerator design since technical guidance in that regard is scarce. In this paper some of the basic concepts on regenerative heat transfer are reviewed so as to provide a rough method of screening for an individual checker format based on thermal effectiveness indexes. Then a characterization of an entire checkerwork by our Regenerator Efficiency Computer Program is made utilizing information of those analytical parameters. Next, regenerator efficiencies and temperatures through the checkerwork height are discussed for refractory applications. Furthermore, a dynamic temperature response of regenerator case walls is presented by coupling RHI computer output data to the transient heat transfer module of FEM Quick Field (QF) Software. Finally, the QF stress distribution module is linked to the temperature distribution to see effects on behavior of refractory structures according to thermal expansion and creep properties.

Reduced Parameters in Regenerator Calculations

The most important aim in heat recovery in regenerators is to achieve the highest preheated air temperature. This depends on the ability of the regenerator to transfer heat from the gas, which is proportional to the product of the convective heat transfer coefficient, α_c, and the specific heating surface, $s.H.S.$ This product of the convective heat transfer coefficient α_c and the specific heating surface $s.H.S.$ is the main parameter in regenerator efficiency and hence the resultant preheated air temperature. That combination happens to appear in the numerators of the mathematical expressions for the so-called reduced parameters such as reduced length and reduced period in the $\Lambda - \Pi$ approach for efficiency calculations. [2,6]

137

By looking at these dimensionless parameters, reduced length in particular, one can see readily how these factors weigh on the thermal performance of a given checker format.

Reduced length Λ
The reduced length, which sometimes is referred to as the thermal size of the regenerator embodies the critical components needed for high heat transfer rates, namely: specific heating surface and convective heat transfer coefficient. Despite the fact that the overall heat transfer coefficient, e.g. radiation and convection heat transfer, from gas to brick is about 5 times as great as that from brick to air [3], the regeneration objective still relies on extracting the heat from the checkerwork to give off to the air. That is why from now on in the reduced length parameter the properties of the gas are to be those of the combustion air

$$\Lambda = \frac{\alpha_c (s.H.S.)}{v_{actual} \rho \; Cp} \times \frac{L}{A}$$

(1)

Where, α_c is the convective heat transfer coefficient [$WK^{-1}m^{-2}$], $s.H.S.$ the specific Heating Surface [m^{-1}], v_{actual} the actual velocity of gas [ms^{-1}], A is the effective cross section of flow [m^2], ρ is the gas density [kgm^{-3}], C_p is the gas heat capacity [$Jkg^{-1}K^{-1}$] and L the characteristic length [m] often referred as the checker height.

Reduced Period Π
This dimensionless time parameter represents the energy storage in the checkerwork. Since brick properties such as density and heat capacity occur in denominator of the following expression the capability of retaining heat increases as long as the reduced period Π decreases.

$$\Pi = \frac{\alpha_{conv} (s.H.S.)}{(1-\varphi)\rho_s C_s} \times P$$

(2)

Here, P is the period [s], $1-\varphi$ the specific brick volume or fraction of brick volume occupation, ρ_s is the brick density [kgm^{-3}], C_s is the brick heat capacity [$Jkg^{-1}K^{-1}$].
Therefore, low reduced period resultant from a high brick density and high heat capacity means a high reservoir of heat in the bricks, and in an ideal situation, namely, an infinitum massive accumulation of energy leads to a zero reduced period ($\Pi>>0$). A further way to utilize these factors is to consider their ratio. The ratio of the reduced period to the reduced length is known as the utilization parameter U, [U = Π/Λ]. The utilization parameter supports the foundation of an efficiency evaluation based on the reduced length and reduced period parameters. A high reduced length and low reduced period will result in a low utilization parameter, which means more potential because there is more room left for the regeneration process.

Indexes of Regenerator Performance
The higher the preheated combustion air temperature, the greater energy savings. In that regard several indexes of performance in regenerators are found elsewhere in the available literature, among them is the regenerator effectiveness ε, which involves only temperatures in the calculation:

$$\varepsilon = \frac{T_{out\,AIR} - T_{in\,AIR}}{T_{in\,FLUE\;GAS} - T_{in\,AIR}}$$

(3)

T_{outAIR} is the temperature of the preheated combustion air leaving top checkers, T_{inAIR} is the temperature of the combustion air below rider arches. $T_{inFLUE\ GAS}$ is the temperature of the flue gas entering the checkerwork.

A second performance index is the Hausen expression for the regenerator efficiency η, which relates the enthalpies that are carried out by the gas streams in cold and hot periods.

$$\eta = \frac{\dot{m}_{air}\, Cp_{air}\left(T_{out\ AIR} - T_{in\ AIR}\right)}{\dot{m}_{flue}\, Cp_{flue} T_{in\ FLUE\ GAS} \; - \dot{m}_{air}\, Cp_{air} T_{in\ AIR}} \tag{4}$$

Where $m\bullet_{air}$ is the mass flow of the combustion air, $m\bullet_{flue}$ is the mass flow of the flue gas, and Cp_{air} and Cp_{flue} are the heat capacity of air and flue gas respectively.

Ideal Regenerator effectiveness
A third index of performance is the Ideal Regenerator Effectiveness. This is based on the reduced length-reduced period approach discussed earlier. It can be used to assess the efficiency of various checker arrangements without the need to run a complete efficiency calculation using special software. The ideal regenerator effectiveness is calculated for several standard checker formats in Table III below. This formula considers zero reduced period, which might be related to an ideal system involving a symmetrical and balanced regenerator having a huge heat storage capacity.

$$\varepsilon_{\Pi \to 0} = \frac{\Lambda}{\Lambda - 2} \tag{5}$$

Checker Brick Data
The basic checker brick features as presented by refractory suppliers correspond to factors that are purely geometrical such as:
-specific heating surface [m^2/m^3] (s.H.S.) or the heating surface related to the volume of the chamber
-effective cross section of flow as a fraction of the overall section, the ratio of the cross section of free passages to the total flow cross section
-fraction of brick volume occupation [m^3/m^3], $(1-\varphi)$, the ratio of checker bricks to the checker volume
In addition to the factors related to brick shape, there are those factors based on material physical properties that embody the thermal diffusivity of the material, such as density, heat capacity and thermal conductivity of the refractory brick.
The last and perhaps the most important factor is the convection heat transfer coefficient, which is obtained mainly by experimentation and dependant on the spatial configuration of a given particular checker arrangement.
The above characteristics play an important role in the thermal performance of the regenerative heat transfer and will be gathered in the reduced parameters for a quick examination for a given checker shape.

Convective Heat Transfer Coefficient
The convective heat transfer coefficient α_{conv} regularly is derived from the Nusselt Number (Nu).

$$\alpha_c = Nu\frac{\lambda}{D} \tag{6}$$

where λ is the thermal conductivity of the gas and D the characteristic length, usually the wet diameter [m]. Since Nusselt formulas for regeneration heat transfer might be proprietary and perhaps controversial among refractory suppliers, the work of Bauer et al, [5] is referred for exemplification. They obtained Nusselt formulas for the six most common checkers used in Glass Industry as shown in Table I. Their experiments were conducted at 300°C and controlling the air velocity at 0.19 ms^{-1} NTP.

Table I. Dimensions of Checkers for Nusselt determination

Experimental Device Area: 0.65X 0.65 m^2 Height: 2 m	flue size D	brick height L	brick thickness B	device heating surface	cross section of channel
Checker Format	[mm]	[mm]	[mm]	[m^2]	[m^2]
Chimney Block Open	140	150	40	0.1739	0.0191
Chimney Block Closed	140	150	40	0.1642	0.0192
Staggered Pigeonhole	140	124	64	0.1584	0.0271
Straight Pigeonhole	140	130	64	0.137	0.0286
Smooth Cruciform	140	420	30	0.2484	0.0195
Corrugated Cruciform	140	420	30	0.2659	0.0195

The Nusselt expression that resulted from their experiments has the following structure:

$$Nu = a(\text{Re}^b + c \cdot Gr)^d \tag{7}$$

Where a, b, c, and d are constants for a given checker arrangement, Re and Gr are the Reynolds and Grashof numbers respectively. The constants for the corresponding checker systems are shown in Table II. Since the aforementioned authors determined parameters at 300°C, related convective heat transfer parameters considering brick surface at 475°C are displayed for comparison.

It is assumed that a 150 K temperature difference is between brick surface and bulk combustion air [3]. The 300-500°C range belongs to the lower checkerwork, in that respect different opinions have been stated regarding that no carved checker surfaces are required in that location but just smooth surfaced shapes.

The differences that exist between various refractory and engineering companies and their calculation of the convective heat transfer coefficient illustrate the complexity and potential problems in modeling the regenerator efficiency.

Table II. Constants for the Nusselt Expression (7) as in [5]

Checker Format	Constants for Nusselt Formula		
	$b = 0.3$		
	a	c	d
Chimney Block Open	1.379	2.856E-06	1.1264
Chimney Block Closed	0.2962	7.03E-06	1.2848
Staggered Pigeonhole	0.01398	0.001248	0.9759
Straight Pigeonhole	0.1715	9.44E-07	2.0256
Smooth Cruciform	0.5174	5.602E-06	1.392
Corrugated Cruciform	0.2727	1.916E-06	1.959

A second illustration of this point is the differences in opinion of the heat exchange process and ideal shapes. One literature reference states the gas film next to the surface brick which is stagnated, is convenient for the heat exchange, as explained with the so-called entry effect that takes place in between canals-rider arches and lower checkerwork [6], whereas others say that less surface exposed to the right angles of the flow is better to avoid condensate build up at that position. Also the more massive format (neither clefts, nor shapes that enable labyrinth flow patterns, for instance) is convenient in that location to enable bearing weight of layers of bricks above them.

Table III shows the reduced length and the ideal regenerator effectiveness for previous formats, which were calculated by using Eqs. 1 through 5.
Table III has been built by considering bulk air temperature of 325°C and normalized air velocity of 0.19 ms^{-1}. It predicts that both Staggered Pigeonhole and vented Chimney block checker would have a good performance in the regenerative heat transfer. Further these same checker arrangements will be analyzed using the RHI Regenerator computer program so as to draw some more information.

Table III. Reduced Length and Regenerator Effectiveness of four common checker shapes according to device size shown in Table II.

Checker Format	Nu	α_{conv}	Λ	ε
Chimney Block Open	25.42	9.15	2.646	0.570
Chimney Block Closed	15.05	5.42	1.270	0.388
Staggered Pigeonhole	31.30	10.82	1.841	0.479
Straight Pigeonhole	22.10	6.86	1.080	0.351

Reduced Parameters in Chimney Block Formats
There are a variety of Solid Chimney formats available from RHI in addition to those analyzed above. Their features are displayed in Table IV.
The reduced length and the effectiveness for a stack of six bricks made from the RHI chimney formats are displayed in Figs. 1 and 2 respectively for combustion air between 250°C and 1000°C. TL 14/15 for solid chimney with vents and TG 14/15 with plain walls turned out the most efficient among chimney block shapes. Actually the combination of TL 14/15 on the upper checker work and TG14/15 for the condensate and rider arches zones has proven to be successful regarding overall performance including mechanical strength, thermal performance as well as in delaying checkerwork clogging.

Table IV. Basic Characteristics of Chimney Block Shapes

TOPFSTEINFORMATE | CHIMNEY BLOCK SHAPES

Shape	Flue size	Brick thickness	Brick height	Brick volume	Pieces	Specific heat transfer area
	mm	mm	mm	dm³	m²	m²/m³
TG 14/175	142	38	175	4.23	88.2	15.9
TG 14/15	142	38	150	3.61	102.9	15.8
TG 15/15	150	30	150	2.88	102.9	17.4
TG 17/175	170	40	175	5.22	64.8	13.9
TL 14/175	142	38	175	3.74	88.2	16.8
TL 14/15	142	38	150	3.11	102.9	16.7
TL 15/15	150	30	150	2.52	102.9	18.9
TL 17/175	170	40	175	4.51	64.8	14.2
TG 150/175	152	38	175	4.48	78.1	15.2
TL 150/175	152	38	175	3.89	78.1	15.9

Brick height / Steinhöhe   Pieces / Stück
Brick thickness / Steinstärke   Shape / Format
Brick volume / Steinvolumen   Specific heat transfer area / Spezifische Heizfläche
Flue size / Kanalweite

The RHI Regenerator Efficiency Software Program

The need of simplification for considering reduced length as a primary parameter to estimate regenerator effectiveness has neglected to take into account the thickness of the brick and the brick thermal conductivity. Our Regenerator Efficiency Software Program, R20, adds these parameters which are found in the reduced period expression to the calculation, and was designed to assist the customer, either a glass manufacturer or a glass engineering design firm, or both in conjunction, in the selection of a checker system most suitable for their application parameters. The program uses two-dimensional, transient, finite-difference heat transfer analysis to develop profiles in the direction of the gas flow and within the plane of the horizontal checker brick course. The effect of checker setting changes on the gross heat input required to melt glass is then calculated.

Checker brick properties (thermal conductivity and heat capacity) are used as a function of temperature to calculate the heat flow in the checker setting. The model with respect to effective surface area, free-plan area, solid volume fraction, and flue size equitably compares checker settings. It recognizes the important operating variables in glass furnace operations and uses well-known techniques to address heat transfer as well as solving the resulting series of descriptive equations.

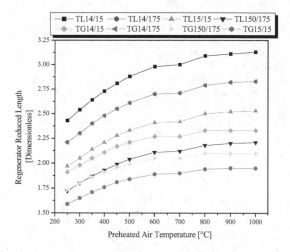

Fig. 1. Reduced length Λ for Chimney Block Formats

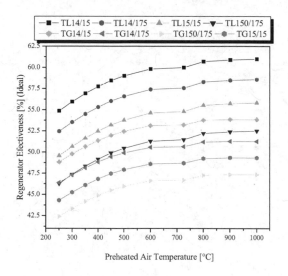

Fig. 2. Regenerator effectiveness ε for Chimney Block Formats

RHI Temperature Distribution Analysis

Figure 3 and 4 shows the temperature charts along the checker work height having a staggered pigeon hole packing. In Fig 3, curves in red and blue represent the brick temperature at the middle of the hot and cold periods respectively. They represent an average of temperatures at the respective height, thus they may be utilized for refractory selection purpose

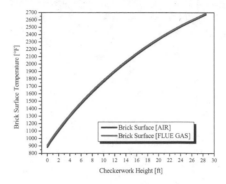

Fig. 3. Refractory Temperature at the Middle of the Regenerative Period

By creating a temperature distribution such as the one shown in Figure 3, the furnace designer can specifically pinpoint critical areas such as the condensate zone of the checkwork (800-1100°C) and zone this area with appropriate refractory materials. This allows for a more cost effective solution by utilizing the higher cost magnesia-zirconia or spinel based materials only in these required areas. It also may allow the furnace designer to make adjustments in design or operations to actually shift the location of the condensate to more desirable/less detrimental locations.

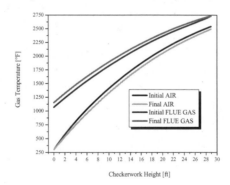

Fig. 4. Gas Temperature at the End and the Beginning of the Regenerative Period

Optimal Reversal Time

One question that is often asked in the regenerator design process is whether increasing the reversal time will improve efficiency [7]. The RHI program is able to point out the drawback of having reversal times beyond 20 minutes. Figure 5 shows that higher temperature for combustion air is achieved with shorter periods, than currently found in some facilities, e.g. 20 minutes instead of 30 minutes.

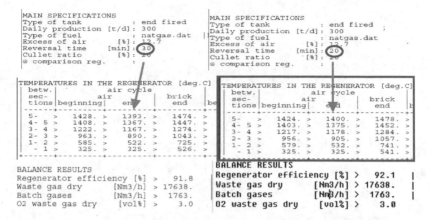

Fig.5. RHI Output Display Comparing Reversal Times

Temperature Distribution of Regenerator Case Walls by Coupling the Transient Heat Transfer Module of FEM Quick Field Software

The regenerator chamber casing plays an important role being both the mechanical support that houses the checkerwork as well as the insulation cover that helps to prevent heat losses. Many efforts have been focused in that past years to model temperature distribution in the checkerwork, whereas attempts to feature the thermal and mechanical behavior in regenerator walls have been rather scarce.

Nowadays, commercial software packages with the capabilities to model various industrial processes are available. Among them is the FEM Quick Field (QF) software from TERA ANALYSIS LTD whose Transient Heat Transfer suite has been applied to feature the regenerator wall thermal response by coupling the RHI regenerator efficiency program output data as input to the QF module.

Linking between RHI Regenerator Efficiency Program and QF Transient Heat Transfer Module

To provide the link between the RHI regenerator program and the QF software, the RHI output temperature field must be tabulated in such a manner that a given time is a cross-reference for the corresponding column data for either the hot period or cold period of the regeneration cycle. From there, a non-linear regression analysis is taken so as to get a time-dependant expression for those temperatures in order to enable their entrance as inner wall border values to the QF software. Eq. 8 shows the specific temperature expression as a function of time as a result of a non-linear regression analysis. An example of this equation is plotted in Fig. 6 where temperatures above the rider arches according to the RHI regenerator program are shown. The shape of the curve of the temperatures plotted in Fig. 6 resembles closely the experimental points obtained by O-I inside glass furnace canals as shown in Fig. 7 as a comparison which provides some literary validation for this model.

$$T_{[K]} = -147926.53 + 148767.68 \cdot \exp\left(\frac{t_{(s)}}{2.96421 \times 10^6}\right) \qquad (8)$$

correlation coefficient $(\rho) = 0.9872$

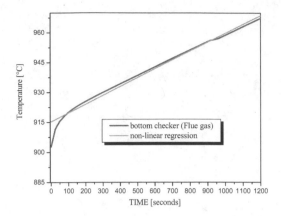

Fig.6 Temperatures at the Bottom Checker as the Hot Period elapses.

Subjecting the temperatures at each checker layer considered in the RHI regenerator program to this same non-linear regression treatment allows the QF input border values to be filled as shown in Fig. 8. Where T_o is normally taken as the bulk gas temperature, but in our case, T_o will approach the brick surface temperature next to the regenerator wall.

Other data to be fed as the border values are:

The convective heat transfer coefficient for outer structures as illustrated in Fig. 9. [3].

The radiative heat transfer view factor between checkerwork bricks and inner walls next to the checkers, which is considered to be equal to one (1) due to the later the value corresponding to infinite parallel planes.

The convective heat transfer coefficient for inner walls has been taken within the range of 4 to 6 [WK^{-1}m^{-2}] for the cold period, depending on the temperature level, and an average of 15 [WK^{-1}m^{-2}] for the hot period.

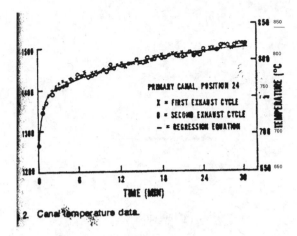

Fig 7.- Chart from: "Data requirements for quantitative analyses of commercial glassmelting system energy performance". Frederick J. Nelson & John D. Novak, Ceramic Bulletin Vol.59, No. 11 (1980) pp. 1142, 1143

Regarding the non-linear behavior of some properties of refractories, such as heat capacity and thermal conductivities, it is certainly taken into account in the QF computations, as well as in the RHI Regenerator program. In Fig. 10 some samples of the QF input charts of the abovementioned properties can be seen.

Fig. 8. Quick Field Transient Temperature Entry Window.

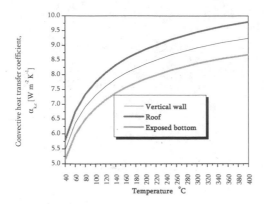

Fig. 9. Outside Walls Convective Heat Transfer Coefficient

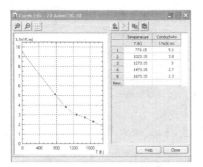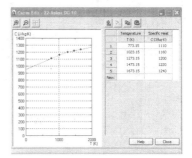

Fig. 10. Quick Field Material Properties Entry Windows.
Left-Hand Side: Thermal Conductivity. Right-Hand Side: Heat Capacity

Dynamic Display of the Temperature Distribution of an End-Port Regenerator Chamber during firing cycles

Once the parameters described above have been entered into the QF code, a dynamic response of temperature for regenerator walls is enabled. Graphic results of QF runs are shown in Fig. 11, where temperature levels are displayed as hues corresponding to the scale at the right-hand side. Timing was fixed according to the entry box shown in Fig 12.

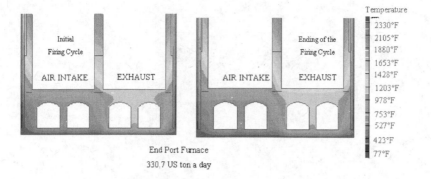

End Port Furnace

330.7 US ton a day

Fig. 11. Temperature Distribution in Lower Regenerator.
Left-Hand Side: Initial Firing Cycle. Right-Hand Side: Final Cycle firing

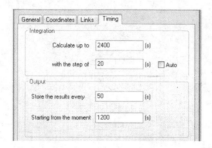

Fig. 12. Time Frame for Transient Temperature Calculation

Typical information related to heat transfer processes could be drawn from this data as well, namely: Local heat fluxes, thermal profiles, temperature gradients, isothermals, average volume heat flux densities, etc., which can be displayed in a variety of charts, curves and figures which include coordinates, integration paths, and so on, for making the thermal analysis easier. As an example, in Fig.13 heat flux distributions taken at the middle of the firing cycle in a regenerator upper structure made out of walls based on Magnesia is shown.

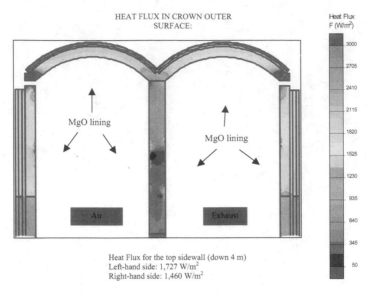

HEAT FLUX IN CROWN OUTER
SURFACE:

Heat Flux for the top sidewall (down 4 m)
Left-hand side: 1,727 W/m^2
Right-hand side: 1,460 W/m^2

Fig. 13. Heat Fluxes in Regenerator Upper Structure at the middle of the Firing Cycle
Magnesia inner lining.

Stress Distribution in Refractory Structures

In addition to the importance of knowing the temperature at every location of the regenerator space, other engineering properties such as mechanical stresses due to the load that bear bricks everywhere in the chamber have to be featured.

By applying the QF Stress Analysis Module to analyze the effect at "room temperature" of the weight exercised by a column above other bricks below on several parts of the regenerator chamber, basic information is obtained to characterize the mechanical behavior of regenerator walls and checkerwork for instance. There are two approaches to tackle stress problems in two dimensions by the QF package, namely, plane stress model and the plane strain model. The latter is formulated by assuming that out-of-plane strains are negligible. Therefore the plane strain model is suitable for structures that are thick in the out-of-plane direction, such as the regenerator center wall being looked at along one of its edges. As far as mechanical properties are concerned, parameters such as elasticity modules and Poisson ratios, are to be fed in the QF database as shown in Fig. 14. Even some arbitrary limits of stress such as cold crushing strength for instance can be considered to envisage failure criteria. Following some examples in that regard are shown:

Fig. 14. Mechanical Properties Input for the QF Stress Analysis Module.

Mechanical Load along Checkerwork Height
In order to emulate a hollow structure, as occurring in a checkerwork, brick density has been multiplied by respective specific volume $[m^3/m^3]$ in the QF input database.

Fig. 15 shows the compression stress distribution σ_{yy} down the 11 m tall checkerwork, for two cases:
1) The staggered pigeon hole packing.
2) The solid chimney with vents at the upper part and with no vents at lower parts.

Chart in Fig. 16 shows the plot of the stresses along checkerwork height. This information is useful for design purposes, for example to evaluate the balance compromise between gaining thermal efficiency, or loosing mechanical stability if a relatively high vertical structure is considered.

Mechanical Load along an End-Port Regenerator Center Wall Height
The center or division wall plays a crucial role in the mechanical integrity of an end port furnace regenerator. As each crown is sprung from the top of the center wall, its skewback bears double the load of each of the outer skewbacks. In addition, the lower center wall has to bear the reaction of the rider arches load at that level. Fig. 17 shows the compression stress distribution σ_{yy} down the center wall. Also a decrease in compression can be seen right below bottom checkerwork since load spreads where there is not a column structure but bearing arches instead then the load increases again when the center wall becomes the canal wall.

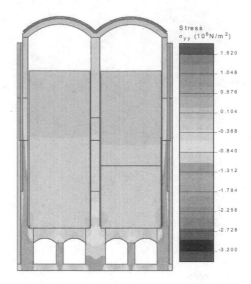

Fig. 15. Stress Distribution in a "Cold" Regenerator Structure.
Left-hand side: Staggered pigeon hole packing checkerwork
Right-hand side: Solid chimney combination: mouse-hole format at the upper part and plain format
at lower parts

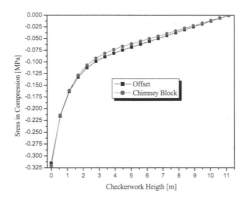

Fig. 16. Stress Distribution along Checkerwork Height according to Fig. 12.

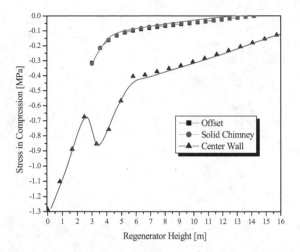

Fig. 17. Stress Distribution along Regenerator Structure Height according to Fig. 16

Coupled Problems in Glass Furnace Regenerators

The effect of thermal strains in Regenerators when heated can be taken into account by the Stress analysis module when temperatures are imported from the QF heat transfer module. This linking is enabled when thermal expansion coefficients have been entered in the corresponding input boxes. Prescribed displacement as border values between structures that meet each other at right angles are fixed in such a way to allow expansion joints thus there will be not undesirable stress built up resulting in fractures.

Besides obtaining the typical output as described in preceding paragraphs, an exaggerated contour is now visualized around the original structure showing a displacement pattern as a result of the refractory thermal expansion. In Fig. 18 can be seen how the deformation is enhanced in upper structures since they are hotter than lower parts of the regenerator as expected. By looking at the outer skewback regions in Fig. 11, it can be seen that there is a gap between the base of the outer skewback and the top of the sidewall to allow for expansion joints. The overlapped contour around the top of the sidewall now meets the base of the skewback resembling the growth of material due to the thermal expansion.

Creep sometimes occurs in bricks when exposed to harsh environments, extremely high loads and temperatures exceeding service specifications often due to an improper selection of refractory. It is interesting to know how far a refractory undergoing such a condition could displace from the original designed structure so as to prevent it from collapsing. Since creep is a phenomenon that occurs after a very long time, starting with a sharp squeezing but decaying in intensity in the long term, predicting subsidence as a function of time is difficult. Hence a hypothetical "negative" thermal expansion coefficient, -4×10^{-5} K^{-1} as exemplification- has been considered in Fig. 19 for emulating deformation in upper sidewalls and upper center wall due to poor hot refractory properties.

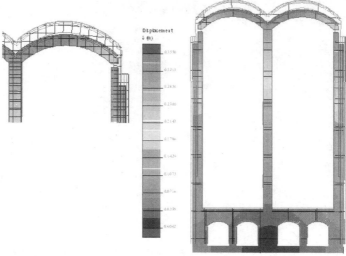

Fig. 18 Displacement of Refractory as a Result of its Thermal Expansion in an Operating Regenerator

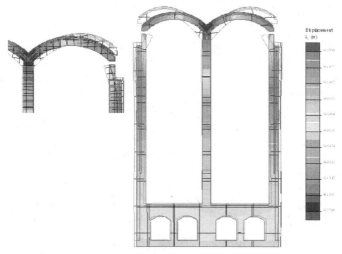

Fig. 19. Emulating refractory subsidence by considering negative thermal expansions in the upper structure lining

SUMMARY

The present report reviewed basic concepts in regenerative heat transfer involved in designing of glass furnace regenerators. These were first applied to obtain a preliminary analysis based on the rough thermal performance among various checker bricks. RHI efforts in that regard have turned out in a homemade Regenerator efficiency software program, which provides more accurate information on the dynamic behavior of the checkerwork. Also by taking the available resources regarding both general purpose engineering software packages and the increased computing capabilities of today, it has been enabled to tackle the regeneration process as a whole by coupling heat transfer and mechanical analysis of the chamber casing so as to achieve to select refractory for longer-lasting and more efficient units. Much of the focus of this work was devoted to set simple guidelines for interpretation and give practical applications on what is usually considered to be part of more complex modeling issues.

REFERENCES

1.- Operation of Counterflow Regenerator, Vol. 4; G.D. Dragutenovic & B.S. Baclic; WIT press. Computational Mechanics Publication, 1998;Southampton, UK and Boston USA

2.- Dynamics of Regenerative Heat Transfer; A. John Willmott; Series in Computational and Physical Process in Mechanical and Thermal Sciences; Taylor & Francis, New York, 2002

3.- Glass Furnaces: Design, Construction and Operation; W. Trier; The Society of Glass Technology, 1983.

4.-Comparison of Regenerators Packing Design; G. Brown; Rockware Glass L.T.D.;
Glass Technology, vol. 26, No. 6, December 1985

5.- Measurement of Convective Heat Transfer for Various Checker Systems; J. Bauer, O. R. Hofmann & S. Giese; Glastechnische Berichte, Glass Sci. Technol. 67 (1994) No. 10, pp 272-279

6.- A Numerical Approach for the Study of Glass Furnace Regenerators; Y. Reboussin, J.F. Fourmiqué, P.H. Marty, O. Citti; Applied Thermal Engineering, 25 (2005), pp 2299-2320

7.-Modeling and Optimal Switching of Regenerators; K.M. Abbott; Glass Technology. 21, (1980), No. 6, pp 284-289.

THINKING GREEN: RECYCLING IN THE REFRACTORY INDUSTRY

Werner Odreitz
RHI AG Technology Center ·
Leoben, Austria

Matt Wheeler
RHI-Monofrax
Batavia, OH

ABSTRACT

As the world's largest producer of refractory materials, RHI recognizes the need to be environmentally conscious as well as to support the efforts of our customers to reduce their carbon footprint. In 2009, RHI recycled more than 70,000 metric tons of used refractory materials received from all markets worldwide. These materials (not all originally manufactured by RHI) are used in a large variety of products and applications with the end goals to save natural resources, reduce carbon emissions created by processing raw materials, reduce landfill, save landfill costs and provide cost savings to our customers. This paper presents RHI recycling philosophy, where we started and what the future holds.

INTRODUCTION

The overall worldwide demand for refractory products per year is in the range of approximately 20 Mio tons. This includes figures from China, which can be uncertain. RHI produces a total of 1,8 Mio tons of refractories per year, having a market share of about 9% worldwide.

The worldwide demand for refractory products in the glass industry is about 600.000 tons per year, excluding China. RHI produces around 90.000 to related to the glass industry, meaning a market share of 15%.

Reclaimed products to be reused in new applications play an important role in RHI philosophy. The amount of recycling products to be put back into production has been consistently rising over the last years as is shown in Figure 1. In 2010 we will receive around 80.000 tons of reclaimed, which equates to a recycling rate of approximately 4,4% compared to the overall refractory production at RHI.

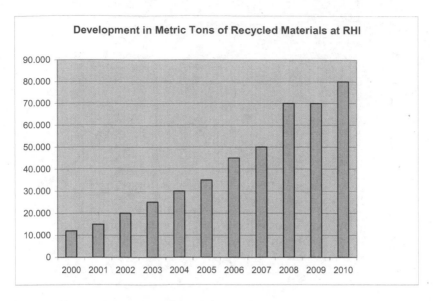

Fig. 1 Development in metric tons of recycled materials at RHI

The main industry sources available to recover reclaimed materials are the steel, glass, ceramic, lime, petrochemical, non-ferrous industries among others. By volumes, the steel industry puts out by far the quantity of reclaimed, due to shorter lifetime of the refractories and continuous re-linings compared to the glass industry.

MOTIVATION FOR RECYCLING

Recycling supports the sustainable protection of natural resources. This means less exploitation and protecting the environment. It also lowers carbon emissions (CO_2 emissions), which are created by producing raw materials. For example MgO is produced by refining $MgCO_3$ into $MgO + CO_2$ (gas). Greenhouse gas emissions are a worldwide focus, recycling supports the efforts to achieve these ambitious goals.

Recycling further reduces landfill and saves landfill costs. Regulations for landfill are getting more and more tight and recycling actively assists these aims.

In an environment moving towards a better quality of life and taking more responsibility to hand over a liveable world to future generations, recycling plays an active role and is part of achieving our targets.

EFFECTS OF RECYCLING

An important consideration and driving force to recycle is to save on raw material costs, especially in these dynamic times of unstable raw material availabilities, volatile prices and changing markets.

Recycling positively influences the environmental balance of a corporation, thus improving the input-output figures considerably.

By taking back reclaimed materials, one is forced to introduce new products in new applications. This widens the product range and offers new opportunities to supply materials in so far unknown or unreachable markets. Recycling is also a big chance for closer cooperation between refractory supplier and customer as we work to solve problems together.

The liaison with recycling companies which do the dismantling work and/or selecting and cleaning of the refractory breakout materials improves further the whole recycling concept and allows us to move from the cherry picking of just a few products to a broader recycling policy and supports the idea of zero waste concepts.

Our customers also have receive advantages from recycling due to lower disposal costs, reduced issues with environmental authorities, less scrap in their backyard, improved environmental figures, and even a financial compensation for certain products. Also, new products based on recycling will be introduced and help save refractory costs. These reasons make it attractive to recycle in a partnership between refractory supplier, recycling companies, and refractory end-users.

RHI HISTORY OF RECYCLING

The RHI recycling concept started many year ago by reusing production scrap into new products. Some production lines always used their internal scrap, while others disposed of them. It took some time and efforts to change the production facility mindset and take away the fear of utilizing secondary raw materials. In times of low raw material costs, protected markets, and high market shares with good margins, the use of secondary materials was not a top priority.

Over the years the market changed, Chinese raw material producers stepped in, prices decreased for a certain time, and the western world began to depend on China for many of the primary raw materials. Production sites were closed in Europe and the United States, the demand for materials like magnesite, fused magnesite, bauxite, graphite and other material could only be met by purchasing from China. At the time it appeared like there were immeasurable amounts of raw materials available. Many Western companies invested in China, set up production sites there or created joint venture companies. At the same time, China's economy started to grow at an unbelievable speed, which finally ended up in a shortage of many raw materials and painful price increases of most raw materials, not only from China, but worldwide. High raw material costs, paired with competition of Chinese final products resulted in many Western companies having serious problems.

RHI did foresee this situation and in the 1990's and started recycling programs in a small scale in Europe to take back Mag Carbon Converter Bricks from the steel industry. These bricks were to be partly reused in Mag Carbon Ladle Bricks and sold back to the customers as recycling brands in lower applications. We faced troubles and invested a lot of research and trials to solve the problems. More and more the idea was appreciated in the market and RHI introduced several recycling brands for steel ladles which performed very well and were designed for special needs of the customers.
Other products like Magnesia Chrome, Alumina Magnesia Carbon, Fireclay, Bauxite, Andalusite and High Alumina followed. Today RHI recycles almost all breakout materials from the steel industry as well as by-products like Sliding Gates, Nozzles, and Well Blocks among others. Together with

designated recycling company partners RHI works out complete recycling concepts for our steel customers.

Refractory Recycling from the glass industry started with AZS materials, which today are still an important part of our recycling circle. Over the years RHI developed concepts for the reuse of different breakout products like Zircon (including ISO press), Zirconia, Dense Chrome, Alumina Chromes, Fused Mullite, Alpha Beta Alumina, Fireclay and Silimanite among others.
More recently, RHI introduced an Integrated Chromium Oxide Circuit (ICOC), which is a recycling program for Alumina Chrome Products. Details of the ICOC program are shown below.

RECYCLING PRODUCTS

Following is a chart of the main refractory products that are recycled by RHI and converted into new products and new applications.

Product	Source	Tonnage per year
Burned Magnesia	Steel, Lime, Others	6,000
Magnesia Carbon	Steel	45,000
Magnesia Chrome	Steel, Others	5,000
Alumina Magnesia Carbon	Steel	8,000
Fireclay	Steel, Glass, Others	3,000
Andalusite, Silimanite	Steel, Glass	2,000
Bauxite	Steel	3,000
High Alumina	Steel, Others	1,000
Zirconia	Glass, Others	300
Zircon	Glass, Others	800
Porcelain	Different	2,000
Fused Silica	Glass, Others	1,500
Dense Chrome	Fibreglass	300
Alumina Chrome	Fibreglass	1,000
Alpha Beta Alumina	Glass	200

Fig.2 Main Recycling Products at RHI

RHI RECYCLING PHILOSOPHY

We started by cherry picking special products either directly from the sources, or together with valued partners, which are actively working at the wreckage site, or are active in selecting, sorting and cleaning refractory breakout materials. The partnership with these recycling companies is important, as we are in the position not only to take single products out of a big scrap pile, but to also offer a comprehensive recycling circle to our customers. Doing this, we are creating a win-win situation for all parties involved, as previously much bigger amounts of leftovers had to be landfilled. Working on comprehensive recycling will close the supply chain circle from mining raw materials until reuse after breakout, so most of the products are sustainable.

Working off breakout piles is not all that easy. Following two examples from steel and glass industry, showing the initial state of the materials after breakout and usable secondary raw materials after sorting these mountains.

Picture 1: Breakout from Steel Industry before sorting

Picture 2: Breakout from Glass Industry before sorting

Picture 3: AZS after sorting and cleaning

Picture 4: MgO-C Scrap after sorting and cleaning

Products and fines which cannot be reused in refractory applications, due to analytical results like high alkalies, out of specification chemical and physical properties, contaminations with steel, slag, glass, or other infiltrations, will be separated and alternative solutions are searched for such as use in metallurgical applications, slag splashing, back filling, covering powders, low cost gunning sprays and so on. Some of the products that cannot be used for refractories will end up in completely different applications outside of the refractory world.

The ultimate goal of recycling is to landfill as little as possible and strive for a zero waste policy. This can only be achieved if the customer, the recycling company and the refractory company work altogether. The customer can help by taking care of the aggregates before and after the last batch, e.g. avoid mixing slag into the breakout, avoid mixing different products into one big pile, take care during draining and so on. It helps a lot, if products from different areas are not mixed, as they have to be selected and sorted later. Taking the timing to handle properly during breakout reduces the overall time, working force, disposable material and costs required to recycle.

RHI INTEGRATED CHROMIUM OXIDE CIRCUIT (ICOC)

Background
Chromium oxide based refractories are today the best material solution to resist the wear conditions in glass fibre furnaces. Unfortunately, under the conditions of glass production such as high temperature, presence of alkalis or earth-alkalis and oxidizing atmosphere, chromium oxide can oxidize to the water-soluble hexavalent chromate. The resulting chromate is toxic and therefore undesirable with respect to the environment as well as occupational health and safety.

Because of its potential for toxic character, chromium oxide containing refractories are subjected to harsh disposal conditions both in Europe and the United States. The EPA Landban regulation (Federal Register-vol.55 no.106-June1, 1990) bans the land disposal of refractories, which have more than 5 mg/l of chrome by the EPA Toxicity Characteristic Leaching Procedure (TCLIP). In Europe, until July 2009 concentration thresholds for disposal were regulated by the individual states. Since August 2009 there are standard limits for deposition in the European community. The current disposal regulations are:

-Total concentration of soluble chromium <0,3mg/l: no restrictions
-Total concentration of soluble chromium 0,3-7,0mg/l: to be disposed on a toxic waste dump
-Total concentration of soluble chromium >7,0mg/l: to be run into the ground

This is determined by the Toxicity Characteristic Leaching Procedure for the concentration measurement as described in the ISO-Standard 10390:1997-05.
The international transport of toxic waste from the generator to the landfill is generally possible and regulated by the Basle Convention.

Position of RHI
As the world's largest refractory producer, RHI offers its clientele a wide range of chrome bearing refractories. A large portion is used in the glass fibre industry. RHI also employs secondary chrome bearing raw material for the production of its chrome bearing products. However, with regard to the quality of the products the use of scrap material is always critical and limited by fluxing agents like silica, alkalis or borate. With the installation of an electrical-arc melting device in its plant in Radenthein, Austria, high quality pre-fused raw materials with consistent quality are produced from different scrap materials. An interesting side effect of the technique is, that the undesirable toxic chromate is reduced to trivalent chromium oxide. This infrastructure allows RHI to offer to the customers the option to recycle a wide range of chrome-bearing products and to put them in a closed chrome material circuit.

ICOC Concept
RHI is able to offer to our customers a very flexible recycling concept for used chrome bearing refractories. This service concerns the transport, the treatment and the reintegration of the material in the production process.
RHI can accept all roughly cleaned (removal of adhering glass and non-chrome containing materials) chrome-bearing refractories ranging from 10 to >95% chrome content.
Usable material is treated and reintegrated into our production. Polluted material or rests are run into the ground. All involved RHI partners as well as the hazardous waste landfill are approved for handling chromium oxide and chromate containing materials. If necessary, a protocol of all treatment steps and a certification of reuse or deposition can be provided.
In Europe, when the glass producer purchases a lining from RHI, the withdrawal service is free of charge from the moment of handover of the material to RHI at the wreckage site. The remaining costs to the glass producer are any costs related to removal, cleaning and sorting the refractory on the construction site according to RHI specifications. RHI assumes the transport to our approved partners, organizes the crushing and handles transport to Austria for fusion. Equally, in Europe RHI is undertaking the disposal of polluted material if necessary. The glass producer will receive documentation from RHI stating the material has been received and either processed for use as a secondary raw material or disposed according to regulations.

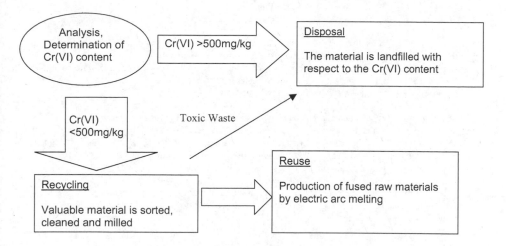

Fig. 3 ICOC Process

Picture 5: Alumina Chrome after sorting and cleaning

CONCLUSION

It is recognised that recycling plays an important role in our daily life. More and more regulations are put in place and it is well understood, that one cannot throw away and dispose of materials as was done in the past. Also disposal of refractory breakout products is regulated, costs for landfill are rising, and some products cannot be disposed without great difficulty and expense. Ideas to recycle and bring back worn materials into the circuit are essential.

RHI has a long history of recycling refractories after lining life in all industries. Recycling concepts are put in place, new products based on reclaimed materials are introduced into the market and the cooperation with recycling partners allows us to work towards zero waste concepts. Advantages include saving natural resources, reduction of carbon emissions, reducing landfill volume, saving landfill costs and ultimately providing cost savings to our customers through all of the above.

Recycling of refractories is one of RHI main focuses for the future in regard to raw material availability and cost savings, as well as to participate in working for a clean, liveable world, which can be passed down to future generations.

RECYCLING OF POST-CONSUMER GLASS: ENERGY SAVINGS, CO_2 EMISSION REDUCTION, EFFECTS ON GLASS QUALITY AND GLASS MELTING

Ruud Beerkens[1]; Goos Kers; Engelbert van Santen
[1]TNO Glass Group, Eindhoven, NL

ABSTRACT

This presentation shows the advantages of re-melting post-consumer glass, but also the potential risks of using contaminated cullet in the raw material batch of glass furnaces (e.g. container glass furnaces).

As an example of potential advantages: increasing the cullet % in the batch of an efficient end-port fired regenerative container glass furnace from 65 up to 75 % decreased the specific energy consumption from 3.95 MJ/kg molten glass to 3.8 MJ/kg and reduced the direct CO_2 emissions with 31 grams per kg glass. Additional, lower indirect CO_2 emissions can be taken into account, since less primary raw materials have to be applied, saving fossil energy in raw material synthesis (e.g. synthetic soda production).

Waste glass has to be sorted and prepared in dedicated cullet recycling (treatment) plants (CTP) to meet the strict quality standards often expressed in maximum mass fraction of ceramics, stones, china, metal (ferro & non-ferro) and color mismatches that can still be accepted. But, also the presence of organic components (fats, oils, sugar, food residues, ..) has to be controlled to avoid production or glass color problems.

Contamination of cullet may lead to:
1. Glass quality problems: inclusions or color changes;
2. Glass melting disturbances by foaming or limited heat transfer into melt;
3. Glass furnace lifetime, by downward drilling of melts of metals, present in the cullet.

The most important problems today are related to the presence of glass ceramics in the post-consumer waste glass and the variable amount of different colors in the glass cullet or fluctuating contamination by organics. Even small pieces of china or glass ceramics in the cullet, with sizes less than 5 mm, may end as glass defects in glass products. Larger pieces of glass ceramics may even lead to severe interruptions in the gob formation process, due to problems with cutting of the gobs with high viscous inclusions. In modern cullet treatment plants most ferro- and non-ferro metals are rather effectively removed.

Glass-ceramics are very difficult to distinguish from normal soda-lime-silica glass, because color and transparency can be almost the same. Specific techniques have to be applied to detect glass-ceramic pieces in the cullet based on X-Ray absorption, X-Ray fluorescence, Hyper-spectral Imaging or UV techniques. Such systems are recently applied in modern waste glass treatment plants in Europe, delivering recycling cullet to the glass industry.

This paper will show the glass defects related to the presence of glass-ceramics and the typical compositions of these inclusions (often present as cord or big knots).

Organic materials can pyrolize within the batch blanket to form carbon rich residues. Carbon or cokes can react with sulfates in the batch and will cause formation of sulfides. A high level of sulfides in the batch (instead of sulfate) will jeopardize the fining process, may cause changes in fining onset temperature, may cause foaming around the batch blanket, or may lead to chemically reduced glass or even amber cords. The process of radiant heat transmission in the melt will change with a variation of the oxidation state of the melt (redox state of batch & cullet).

Examples of glass defects caused by **metal contamination** in the batch will be shown. The composition of the metal inclusions (glass defects) in the glass product may be very different from the composition of the original contamination, causing the defect. An example is metallic aluminum that leads to silicon inclusions in the glass. Nickel sulfide can be formed by pollution of the glass melt by stainless steel flakes. Liquid metals are very aggressive towards the refractory bottom materials of the tank. A droplet of molten lead for instance will drill a hole in the refractory layers: "downward drilling".

Recycling of Post-Consumer Glass

The presentation will show the relation between glass defects and their origin in contaminated cullet and the mechanism of defect formation or defect conversion.

1. RECYCLING IN THE GLASS INDUSTRY

1.1 GENERAL INTRODUCTION

In almost all glass production processes, the raw material batch contains some broken glass. This broken glass may originate from own production, due to glass product rejects or other glass masses that cannot be brought on the market. Another portion/fraction of glass to be recycled in glass production originates from glass processing plants (as waste from cutting losses for instance) or it is glass from replacement of old returnable bottles by new bottles. Generally, this glass is hardly polluted and has a well-controlled composition. However, most waste glass recycled in the container glass and glass wool industry is post-consumer cullet [1-4]. The waste glass is often collected by the municipalities by using large collection banks. Sometimes, the banks are separated in different sections for different glass colors. This post-consumer glass is subject to a high level of contamination, for instance by paper, plastics, organics (food residues, sugars, fats), metals (lids), ceramics or stone and china or glass-ceramics (cookware, oven windows).

Figure 1   Photograph of oven with glass-ceramics windows.
Glass-ceramics compositions show typically a high Al_2O_3 (20-25 mass %) content.
Their melting processes require much higher temperatures compared to soda-lime-silica glass.

The broken glass, to be recycled in the glass melting process is called recycling cullet. We distinguish:
- Internal cullet: almost same composition as glass to be produced; cullet originates from reject during glass production or from transition glass during color changes;
- External cullet: this cullet can originate from post-consumer glass or from glass processing plants.

Since recycling cullet is a very important raw material, used in large quantities in container and glass wool production, the specifications of the required composition and purity of the recycling cullet are very strict. Out-of-spec cullet may cause problems with contamination by organics (will cause glass oxidation fluctuation and color variations, foaming, odor, color streaks), metals (inclusions and stress in glass surrounding inclusions, drilling of metal droplets through melting-end bottom), stones/ceramics/china (inclusions and cords in glass product, problems with melting and forming process).

Post-consumer waste glass is collected at many sites in villages and cities (up to 1 collection bank / collection point per 650-750 citizens in the Netherlands). From there, the collected waste glass is transported to dedicated cullet treatment plants (CTP) to process the cullet to the required sizes and to remove most of the contaminants in order to meet the specifications that are developed by the glass producer, which recycles the glass in the glass melting process. Very regularly the treated cullet has to be checked on level of contamination, not only the mass fractions of pollutants in the cullet, but also the composition and sizes of the pollutants are important. Very small ceramic pieces may dissolve in the glass melt, but china or glass-ceramic pieces larger than few millimeters will not completely dissolve (melt) and cause defects in the glass products. The level of glass recycling in the raw material batch is highest in the container glass sector and glass wool production.

In the melting processes for flat glass, special glass or tableware glass (domestic glass), generally only own cullet is used or cullet from downstream glass processing facilities. Post-consumer glass is hardly used in these sectors, because of risks of contamination of the glass melt. However in the future, when waste glass sorting technologies are improved, post-consumer cullet recycling may become of increasing importance in flat glass production. Techniques are developed to process waste glass from automotive sector (recycling cullet processed to be suitable for flat glass or glass wool production) or to process the waste glass in such a way, that impurities will not harm the melt process or glass quality. Especially in flat glass production contamination of recycling cullet by steel (nickel containing steels/alloys) should be avoided at all times, to avoid formation of nickel sulfide inclusions.

In fibre glass production, the waste (rejected product) is called scrap. This scrap contains organic binders or finish and the material may be very difficult to handle because of its fibrous nature. The removal of the organic coatings and the processing of the scrap becomes an increasingly applied technology to generate a material suitable for internal recycling.

1.2 CONTAINER GLASS PRODUCTION & RECYCLING

Some of the collected waste glass in European countries is exported to other countries and there is also import of external recycling glass. A large part of the produced glass in some countries is exported with the packaged products (wine, beer, fruits) and this glass will hardly return as waste glass for recycling in the same countries in the same glass production sites.

The production of different types and colors of glass in a country often does not reflect the use of the glass in the same country. The beverage, wine or beers packaged in glass are often exported to other countries or even overseas. The post-consumer glass is collected (or disposed) in other countries than where the production of this type / color of glass take place. Countries with high levels of green glass in the collection banks, but with mainly glass furnaces producing flint (colorless) glass may be not able to recycle all the cullet, example is the UK.

For flint (non-colored glass) glass production, the recycling cullet should be free of colored glass. There can be a net export of glass from one country to another part of the world and the mass of glass returning in the collection banks in this country is much less than the production level. In these cases cullet is imported (such as in the Netherlands) or the cullet recycling levels are limited.

For manufacturing of green container glass, multi-colored (mixed cullet: not separated on color) recycling cullet ("mixed cullet") can be used in very high quantities (> 80 % in the batch). For the production of amber glass, only a lower percentage of mixed cullet can be used, due to the chromium oxide content in the green glass fraction of the mixed cullet, but for flint glass no mixed cullet can be used at all. Especially chromium oxide in the cullet will lead to problems in flint glass and at high quantities in amber glass production, because it gives the glass a strong green color.

In 2008, the European Union showed a recycling ratio of 65 % collected waste container glass versus consumed container glass [1]. There is a very intense transfer of recycling cullet between EU member states. In 2008, almost 11 million metric tons of waste (post-consumer) glass has been collected in the 27 EU countries and this quantity is almost completely recycled. For instance in 2008, the import of recycling cullet into the Netherlands was estimated on 300.000 tons glass, excluding flat glass cullet used in glass wool production. Total production of container glass in the Netherlands, about 1.1 million metric tons.

In order to increase the amount of recycling cullet for manufacturing flint and amber glass (up to 80 - 90 %) much attention has been spent in the separation by color at the source (triple glass banks) or to sort on color afterwards with optic-mechanical color separators.

Figure 2 shows the development of glass recycling and color separation in the period since 1990 in the Netherlands for example: in 1995 the percentage of recycling glass separated by color amounted to 52.9 %. By the year 2008, more than 60 % of the waste glass in many European countries (e.g. the Netherlands) is sorted on color at the source (collection banks).

For the production of colored container glass, flat glass cullet can be used. Soda-lime-silica flat (float) glass shows a similar composition compared to container glass, however iron content may be different and generally the soda (Na_2O) concentration in float glass is higher than in container glass. Because of the relatively high iron oxide contents in most float glasses (> 0.08 %), this float glass cullet is not suitable for flint glass production or only at small additions. Float glass can be used in raw material batches for green container glass, amber container glass and glass wool production.

In principle, a further increase of color separation is not needed, because sufficient capacity is available in the Dutch glass industry [2, 3] for using the residual multi-colored cullet for the production of green and amber container glass. In most EU container glass industries, glass recycling has been applied to such an extent that recycling cullet now has become the main ingredient in the batch.

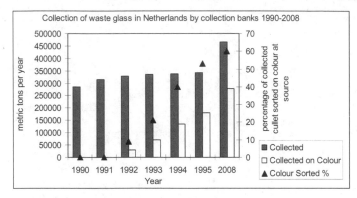

Figure 2   Development of glass recycling in the Netherlands, sources: FEVE [1] and "Duurzaam Verpakkingsglas". Between 1990 and 2008 the number of municipal collection   banks for waste glass increased from 15.000 to 25.000 in the Netherlands.

In South-European countries, recycling of container glass also has become very important both economically and ecologically. The level of glass recycling or the glass recycling ratio in a country or geographical area depends not only on the possibility to replace part of the normal raw materials by cullet, but it is mainly dependent on the cullet collection logistics, the number density of post-consumer glass collection banks, the population density, the presence of dedicated cullet treatment plants and the balance between import and export intensities of packaging glass of different colors. For instance, in the UK, there is a surplus of green cullet, but a lack of sorted flint glass recycling cullet.

In table 1, a survey is presented of the absolute and the relative amounts of disposable glass recycled in the different European countries during 2008. Some countries use a very effective collection system and show a good balance between glass production and glass collection, or even import more glass than can be recycled. Examples are Sweden, Finland, Austria, Switzerland, and Denmark.

Special or lead crystal glass or glass ceramics may enter the glass recycling streams to be used for container glass furnaces. This may cause melting problems or contamination of the soda-lime-silica glass with metals such as lead [4]. According to the European packaging directive the packaging material may not contain the metals: lead (Pb) plus mercury (Hg) plus cadmium (Cd) plus chromium-VI (Cr^{6+}) above a certain limit value. For container glass this is set on 200 mg/kg glass. Glass crystal articles may enter the cullet recycling scheme, because these types of glass are also disposed in the municipality glass collection banks. The lead contents in the produced soda-lime-silica glass will increase by increased recycling of glass cullet contaminated with lead crystal glass. The recycling cullet may contain up to 200 mg lead per kg.

Table 1   FEVE data [1] of container glass consumption and collected waste glass in the EU countries and neighboring countries in the year 2008.

COUNTRY	National Consumption (thousands tons)	Metric tons collected (thousands tons)	Recycling rate %	Official Source
AUSTRIA	268	224,300	84	FEVE
BELGIUM	310,248	297,312	96	Estimated
BULGARIA	160,000	36,000	23	FEVE - Provisional
CZECH REPUBLIC	228,936	140,870	62	FEVE
CYPRUS	18,684	1,950	10	Eurostat - 2007
DENMARK	142,400	124,900	88	FEVE Source - 2007
ESTONIA	22,500	7,238	32	FEVE
FINLAND	60,645	55,545	92	recycling - Actually recycled - 80%
FRANCE	3.200,000	1.960,000	61	FEVE
GERMANY	3.122,065	2.545,441	82	FEVE
GREECE	201,000	24,000	12	FEVE - Provisional
HUNGARY	174,000	42,300	24	FEVE
IRELAND	158,000	127,000	80	FEVE
ITALY	2.138,825	1.540,000	72	FEVE - Collected for recycling - Actually recycled - 65%
LATVIA	68,317	24,188	35	Eurostat - 2007
LITHUANIA	84,069	30,287	36	Eurostat - 2007
LUXEMBOURG	27,111	20,240	75	Eurostat - 2007
NETHERLANDS	572,000	461,000	81	FEVE - 2007
POLAND	777,542	279,959	36	Eurostat - 2007
PORTUGAL	431,500	223,430	52	FEVE
ROMANIA	206,000	22,000	11	FEVE - Provisional
SLOVAKIA	137,410	51,788	38	FEVE
SPAIN	1.613,000	972,658	60	FEVE
SWEDEN	186,000	174,000	94	FEVE
UNITED KINGDOM	2.630,000	1.613,310	61	FEVE - 'Actually recycled' glass - Collected for recycling is not available
TOTAL EU 27*	16.938,052	10.999,716	65	
NORWAY	61,868	56,995	92	FEVE
SWITZERLAND	343,210	325,624	95	FEVE
TURKEY	477,200	92,340	19	FEVE
TOTAL EUROPE	17.820,330	11.474,675	64	

1.3 RECYCLING OF FLAT GLASS AND SPECIAL GLASS TYPES

Until now, the application of (external & internal) recycling cullet in the manufacture of flat glass is still limited to a maximum of about 20 to 30 % of the total batch due to the still relatively low amount of supplied flat glass cullet with required quality. There is no fundamental reason that flint container glass cullet cannot be used in float glass production, but the cullet should be clean and stable in its composition.

The largest supply of flat glass cullet originates from the reprocessing of automotive glass or from processing waste of flat glass in the manufacture of glass windows panes or thermal toughening processes. In Germany several installations are in use for processing different types of flat glass waste automatically, like double glazing and laminated automotive windows. In this equipment crushers separate flat glass from the other components (PVB-foil, double glazing spacers, etc.). The recycling of flat glass is anticipated to become economically more attractive in the future and therefore will continue to increase. Because of strict environmental legislations in

Europe, the recycling of special glass types will also become an important issue in the future.

2. SAVINGS & ENVIRONMENTAL BENEFITS OF GLASS RECYCLING IN GLASS PRODUCTION

Recycling of **external glass waste** has been implemented mainly in the container glass industry. The obtainable savings in raw materials and energy indicated here, specifically apply for the manufacture of glass containers (bottles, jars). However, the same trends or general observations will also apply for glass recycling in other glass industry sectors such as in glass wool production.

2.1 SAVINGS IN ENERGY CONSUMPTION [4-7]

The total energy consumption for the production of container glass can be divided in the energy requirement for:

1. Extraction, mining or synthesis, processing and transportation of raw materials:
   Especially the production of synthetic raw materials is often associated with energy intensive processes, thus showing high indirect energy consumption.
   Replacing a part of the normal batch, containing these synthetic raw materials, by glass cullet will result in indirect energy savings.
2. Melting-in and fining process (energy consumption of the glass furnaces).
3. Forming, cooling and post-treatment (applying coatings, inspection, packaging).

By increasing the amount of foreign (external recycling) cullet in the raw materials, energy can be saved in the first two indicated areas: the raw material processing (indirect energy savings) and the melting process (direct energy savings).

Raw materials

In the processing of the raw materials, the production of synthetic soda ash from sodium chloride brine is the most energy intensive process: about 9-13 MJ/kg soda [5-7]. The exploitation, mining, transport and refining of trona based soda (natural soda) will consume about 7-8 MJ/kg soda. The primary raw material batch (normal batch) for soda-lime-silica glass production contains about 20-23% soda ash. The treatment of cullet requires much less energy, about 0.18 MJ/kg cullet. Thus, each kg of the primary raw materials replaced by cullet saves 1.9-2.35 MJ per kg glass, mainly due to the reduced soda usage.

Melting and fining process

In re-melting cullet, no energy is used in a glass furnace for endothermic fusion and carbonate decomposition reactions and energy is not required anymore to heat up the batch gases (mainly CO_2), that will be absent when using only cullet. Decomposition of carbonates in the raw material batches upon heating of the batch in the glass furnace requires high levels of energy. Without these carbonates this energy is saved. Therefore the reaction energy required, decreases as a function of the amount of cullet in the raw materials.

The fuel consumption reduces by about 2.5 % per 10% increase in the cullet ratio (10 % more glass produced from cullet instead from primary batch components), purely calculated as melting energy saved. However, in practice it appears that the melting process runs faster by using more cullet and the melting capacity in the same furnace can be increased. Energy balance model studies and industrial data show that specific energy consumption will decrease in most cases when increasing the pull of a glass melting furnace as far as the thermal limits of the furnace have not been reached.

This combination of increased cullet ratio and pull will result in net energy savings in the total energy of 2.5 - 3% per 10 mass% replacement of normal batch by recycled glass [2, 3, 4, 7, 8, 9].

In figure 3, a schematic survey is presented of the specific energy consumption as a function of the percentage of cullet used. Figure 4 shows the effect of increased specific pull of a typical container glass furnace on the specific energy consumption.

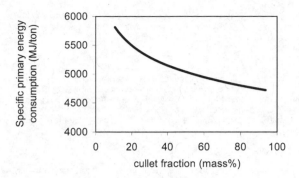

Figure 3 Schematic example of the required energy for melting 1 ton of glass as a function of the percentage cullet according to benchmark analysis of a large set (±120) of industrial container glass furnaces (TNO Benchmarking 1999).

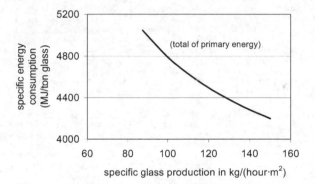

Figure 4 Effect of increased pull on specific energy consumption of a regenerative container glass furnace (example)

By increasing the cullet ratio in the glass forming raw material batch from 10% to 80 mass% (on glass basis), the required (thermal) energy for the **preparation of raw materials** is reduced from 3.5 to 1 MJ/kg (about 70% energy savings), while the required melting energy (energy consumption glass furnace), for an example of a typical container glass furnace, is reduced from about 5 to about 4.1 MJ/kg. But, for colored container glasses, where often large quantities of recycling cullet are applied, more electric boosting has to be applied, because the radiant heat from the combustion chamber will be poorly transmitted by the colored glass melt, about 0.2-0.4 MJ extra electric energy per kg glass has to be provided by electric boosting, but less fossil energy is used. Total primary energy consumption is almost independent on glass color.

Effect on greenhouse gas emissions
Because of direct (in glass factory) and indirect (supply chain of raw materials) energy savings, especially fossil fuel energy savings, the CO_2 emissions will be decreased by using increasing amounts of recycling cullet.

For example an average natural gas fired container glass furnace will emit about 485 kg CO_2 per metric ton molten glass, when using normal batch (primary raw materials) and only 225 kg CO_2 per metric ton melt when using only cullet.

2.2 SAVINGS IN RAW MATERIALS

Besides energy savings and reduction of the amount of domestic waste, the re-melting of recycling glass also offers remarkable savings of primary raw materials: by using 100 kg cullet about 118-122 kg of primary raw material is saved. The difference of 20% is a consequence of the fusion losses (decomposition gases) during melting from primary raw materials like soda, limestone and dolomite. For float glass about 119-122 kg of dry normal batch is needed for 100 kg glass, for most container glass compositions only about 118-120 kg dry normal batch is required for 100 kg of glass, due to the lower amounts of soda in case of container glass.

2.3 EFFECTS OF CULLET RECYCLING ON FLUE GAS EMISSIONS [5]

By melting with a high percentage of recycling cullet, the emissions of most flue gas components will be reduced, but for some components in the flue gases, the concentrations may increase:
• The amount of NO_x emitted will decrease because per kg glass produced less fuel is required and the furnace temperature may be reduced when increasing the cullet ratio in the glass forming batch. For air-fuel oil or air-gas fired glass furnaces the NO_x-formation in the flames mainly depends on the temperature level and oxygen contents in the core of the flames;
• The amount of CO_2 emitted will decrease, because less fuel is required and because in the raw material mixture less carbonates are present when decreasing the normal batch fraction. For a 100% cullet batch, the CO_2 emissions per ton molten container glass are in the order of 225 kg/metric ton melt for natural gas firing, for a normal batch this is 480-500 kg CO_2/metric ton molten glass;
• If the amount of sodium sulfate (Na_2SO_4) as a fining agent can be reduced, the SO_x-emissions will be reduced considerably. However often a high amount of sodium sulfate is used for oxidizing the recycling glass when containing high levels of reducing organic contaminants. Then no or hardly any reduction of the SO_x-emissions will occur. The potential reduction of SO_x will be higher in fuel-oil fired furnaces when increasing the cullet level in batch, because about 25-30 % less fuel oil is required to melt glass from cullet, compared to melting from normal batch. Thus, the sulfur oxide emissions caused by the sulfur in the fuel will be considerably reduced in that case.
• Provided that the cullet itself does not contain too much fluorides or chlorides, the emissions of these components are decreased when the furnace temperature is decreased. At a lower temperature of the melt surface, less fluorides and chlorides will evaporate from the melt. In addition less soda ash is used, which also contain chloride impurities, in case of the application of synthetic soda qualities. However, waste glass cullet may contain high concentrations of chlorides (PVC or other polymer contamination, or glass with high chloride content) and fluorides, in that case HCl respectively HF emissions will increase by applying this cullet in the furnace.
• By application of lower furnace temperatures and reduced additions of fining agents, the dust emissions can be reduced, because fewer components will evaporate from the melt, especially sodium components. A main condition for a net reduction of the dust emissions is that the recycling glass should not contain very fine glass powder, because this will be blown off easily. Typically, about 10% of the flue gas dust of soda-lime-silica glass furnaces is formed by carryover of fine particulates from the raw materials.
• Heavy metals: cullet may contain some lead crystal glass or other types of glasses that contain heavy metals or other components of environmental concern such as antimony or arsenic. Lead, arsenic and antimony compounds in the glass can partly evaporate upon melting. Thus depending on the source of the recycling cullet, the emissions of lead (Pb) may increase. But most of these components are in particulate form at flue gas temperatures, and can be removed from the flue gases by filtration (bag filters or electrostatic precipitators).

2.4 CULLET PREHEATING WITH FLUE GASES

The preheating of the raw materials, including the cullet, before charging them into the furnace, using the heat of the hot flue gases exiting from recuperator(s) or regenerators will result in considerable energy savings for the melting process.

Energy savings of 12 up to about 18 % can be achieved by preheating raw material batch and cullet up to 300 °C. This will in all cases result in specific energy savings (energy per kg molten glass) of at least 12 % when preheating all (complete) raw material plus cullet.

Some pre-heating systems are designed for cullet preheating only and these systems are preferably used for glass furnaces operating with more than 70 % of cullet.

A few preheating devices can handle complete batches, but these batches need to have at least a certain percentage of glass cullet. Today (2010), there are no systems available that can preheat normal batch by flue gas heat contents without cullet in the batch.

3. QUALITY ASPECTS OF RECYCLING GLASS

The quality of recycling cullet is mainly determined by the presence of contaminants in the delivered recycling glass and the composition of the glass pieces [2, 4, 8, 9]. The most common contaminants or other compounds in collected waste glass and other aspects include:

a. Impurities:
   - ceramic, stone, porcelain/chinaware;
   - ferro and non-ferro metals;
   - special glass types (like quartz = vitreous silica glass, glass ceramics, lead glass, opal glass);
   - organic waste: paper, plastic, sugar, fats (food residues).

b. Color composition of the cullet

c. Moisture content

Furthermore, the **cullet sizes (size-distribution)** are important.

These aspects will be discussed in the following sub-sections. Subsequently attention will be given to the characterization and control of the redox state of recycling glass.

In cullet treatment plants (CTP) most contaminants should be separated from the waste glass to make the cullet suitable as a raw material in glass production. Contamination of waste glass can be caused by several steps in the waste glass collection and treatment scheme:

- Disposal of non-glass articles in municipal glass collection banks;
- Disposal of non soda-lime-silica glass types, such as lead crystal or glass ceramics in these collection banks;
- Pollution of the transport systems of collected cullet: transport from collection banks to cullet treatment plants or to glass factory;
- Collection of industrial cullet (from glass processing, glass cutting) with non soda-lime-silica fractions and contamination of soda-lime-silica glass streams with this cullet;
- Contamination during cullet treatment in CTP: breaking, sieving, sorting on color, sorting of contaminants. Steel contamination by abrasion of equipment may take place
- Contamination during cullet storage, especially in open spaces not separated by walls or sheltered by roofs. Cullet can become contaminated with pieces of wood, dead leaves, concrete pieces, plastics, other glass types, etcetera, when the cullet is not stored in dedicated and clean areas.

3.1 IMPURITIES

The presence of foreign components (contaminants) in recycling glass can cause severe problems in the glass manufacture processes. The most harmful materials are ceramic materials like stones and chinaware, being hardly fusible or which hardly dissolve in the molten glass, and non-ferro metals, especially aluminum (metallic Al). Alumina particles (Al_2O_3) larger than 1 to 1.5 mm or china/ceramic/glass ceramic pieces larger than 2.5 – 3 mm might not completely dissolve in the glass, melted in an industrial glass furnace. Non-melted / not completely dissolved ceramic particles introduced in the glass product through the recycling of polluted cullet may cause fracture of these glass products and therefore these glass products will be rejected by quality control systems during inspection. Larger glass ceramics inclusions or highly viscous knots in the molten glass can damage the shear blades of the gobbing system. Furthermore it can cause interruption of the gob delivery process to the forming machines and results in production interruptions.

Metallic aluminum pieces (Al) in the batch or glass melt reduce SiO_2 into Si-globules (Si-spherically shaped while molten, due to the high interface tension), which have a different expansion coefficient, compared to the surrounding glass. These inclusions may lead to fracture caused by thermal stress developed during cooling (by the difference in the thermal expansion of the silicon compared to the surrounding glass).

Metallic lead impurities or reduced lead will sink to the tank bottom and there it will attack the refractory ("downward drilling"). Lead oxide containing glass in the cullet (such as lead crystal glass contamination) in combination with organic materials or metals in the cullet can cause the formation of metallic lead upon heating of the batch, with danger of downward drilling of the melt tank bottom.

Some reactions of different metal contaminants entering the glass furnace:

Iron (or iron in steels):
Iron reduces sulfates, leading to amber chromophore cord:

$$4Fe + SO_4^{2-} \rightarrow 4Fe^{2+}/Fe^{3+} + S^{2-} + 4O^{2-}$$

Lead (metallic lead):
Liquid lead metal (density 11.3 kg/dm^3) drills into refractory bottom

Copper:
Forms copper streak or Cu-inclusions in glass

Nickel or stainless steels flakes: may form very small NiS inclusions (huge problem for float glass production for thermally toughened glass)

Aluminum:
Reduces silica sand: $4\,Al + 3\,SiO_2 \rightarrow 3\,Si + 2\,Al_2O_3$ **metallic** Si forms silicon spheres

Aluminum impurities can be removed to a large extent with electromagnetic sorting machines (eddy-current systems).

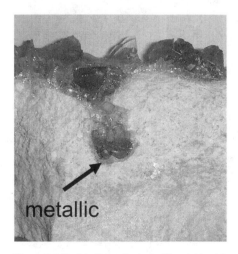

Figure 5a Tank bottom refractory with metal-lead down drill

In processing recycling cullet practically 100% of the ferro-metals can be removed with magnetic separators unless the metals stick to the glass pieces.

Other glass types like **vitreous silica** and **glass ceramics**, generally occur as impurity, and create stress-concentrations within the glass product. Glass-ceramics are melted at much higher temperatures compared to soda-lime-silica glass and viscosity is at least one order of magnitude higher at same temperature. Most glass ceramics contain high concentrations of alumina and are using antimony, arsenic or tin oxides as fining agents. These types of impurities are difficult to separate in processing the recycling glass. Vitreous silica and glass ceramics visually resemble soda-lime-silica glass and have similar electric properties as normal glass. Identification of these types of contaminating glasses is based on techniques that are sensitive for glass composition differences, such as UV absorption characteristics or X-Ray absorption or fluorescence. Other spectral methods that can distinguish between soda-lime-silica glass and other glass types are under development.

Figure 5b Glass with inclusions caused by glass ceramics in cullet

The **organic components** in the recycling glass [2, 3] can only be separated to a certain extent (organic materials can be washed off, or partly be eliminated by fermentation processes or pyrolised / burnt off at high preheat temperatures). These organic impurities have a strong reducing effect on the glass melt. Consequently the oxidation state of the melt is decreased (Fe^{2+}/Fe^{3+} concentration ratio increases), which may also cause foaming problems (extra sulfate has to be added as oxidant to the batch), fining instabilities and color variations in the glass.

Some of the organic impurities (paper, plastics) can be extracted by continuous air suction above the contaminated cullet stream, the largest part of the relatively light papers and light plastics are separated by this way (air sifting).

Many glass recycling plants partly remove stones and other ceramic particles manually. In addition, also automatic opto-mechanical separation systems are applied to remove the larger ceramic particles (> 8 mm). Little stones, chinaware and other ceramic particles smaller than about 8-10 mm are still difficult to remove automatically. The lower size detection and sorting limits are about pieces of 8 mm.

Depending on the melt conditions, ceramic or glass-ceramic particles in the cullet with dimensions between about 2 and 8 mm will lead to inclusions in the glass products with the harmful consequences as indicated. The critical sizes depend on the type of ceramic stones or glass ceramics and the residence time, temperatures and melt flows in the glass melting tank. Some companies apply grinding of all cullet, that is suspected to be contaminated with glass ceramics, china and stones, into particles smaller than about 1 to 2 mm. The pulverized ceramic impurities (< 1-2 mm) then will be able to dissolve into the glass melt. However, this grinding will increase the energy consumption of the cullet processing. Moreover, it generates a larger fraction of very fine cullet powder, prone to carry-over when charging the batch in the glass furnace. Fine cullet also adsorbs large quantities of water, and organic contamination is often concentrated in the finer cullet fractions.

The preferred route is to separate organic components and metals from the cullet prior to grinding and to avoid storage of the grinded cullet in humid environments. It is preferred to grind clean cullet instead of strongly organic material contaminated glass cullet. Grinding of fresh cullet (not washed and without fermentation of the organic material) leads to cullet powder with a high reducing power. The organic material in fine cullet will hardly undergo fermentation because of limited oxygen penetration in fine cullet piles.

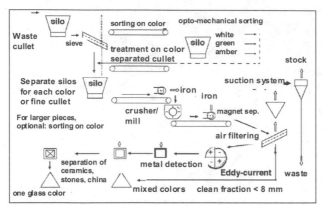

Figure 6   Flow sheet of glass recycling plant [10] (without a sorting system for glass-ceramics separation)

In figure 6, a flow sheet is presented of a state-of-the-art processing plant (Cullet Processing Plant) for obtaining recycling glass as raw material [10]. Sorting on color will generally be applied for very coarse cullet. In many countries sorting of different glass colors at source is applied by providing different collection banks for different glass colors. Color sorting and cleaning/fermentation is preferred, before crushing of the glass. By sieving, a separation on size takes place first: the fractions of large cullet pieces can be separated on color by opto-mechanical sorting, and subsequently the each color of cullet fractions can further be treated separately.

For mixed (different colors) cullet or color separated coarse cullet, separation on ferro-metallic pieces will be applied. Then grinding/ crushing of the coarse cullet takes place and again ferro-magnetic separators will be used to capture residual ferro metallic components. A part of the paper, plastics, foils or other low density materials will be separated by an air suction system (air sifting). A sieve separates the fine from the coarse cullet, occasionally the coarse pieces might be sorted on color.

The non-ferro metallic components will be separated from the color sorted cullet or from the mixed coarse cullet (8-25 mm) by eddy-current devices and the ceramic, china pieces or stones (> about 6-8 mm) are separated by opto-mechanic systems, based on absorption of laser light by non-transparent pieces. These systems generally cannot identify or detect glass-ceramics or non soda-lime-silica glass types. Glass ceramic pieces larger than 8-10 mm can be detected and sorted out with systems developed in the last decade. The detection of glass-ceramics, to distinguish this material from 'normal' glass is based on a difference of X-Ray absorption, X-Ray fluorescence or UV absorption or other spectral properties determined by glass composition. The fine cullet (< 1.5 mm) will be used for glass production, often after additional separation of most of the metallic contaminants, together with the treated coarse cullet.

3.2 COLOR COMPOSITION

The color composition of recycling glass after cullet processing largely determines the usability for different glass production process. Non-color-sorted ("multicolored" or "mixed") cullet can be used for more than 80% of the total batch in melting green glass. For melting brown (amber) glass multi-colored cullet can only be used up to 60% of the batch (more than 60 % mixed cullet would give high chromium concentrations coloring the glass green, which destabilizes the amber coloring), dependent on the color composition of the mixed cullet. A large quantity of green glass in the mixed cullet will give the amber glass a greenish shade (due to the introduction of large concentrations of Cr^{3+}) and this will decrease the amber purity of the glass and lowers the acceptance by the customers.

Multi-colored cullet cannot be used at all for manufacturing high quality white (flint) glass: according to some specifications only less than 50 g of colored cullet per metric ton is allowed in recycling glass to be used for the manufacture of pure flint glass. These limitations originate from the presence of the colored ions Fe^{2+} and Cr^{3+} in green and amber glass cullet.

Because the different colored glasses have different oxidation states, the color composition of multi-colored cullet will also have an effect on the oxidation state of the obtained melt and the color of the resulting glass product (because the oxidation state influences the redox ratio: Fe^{2+}/Fe^{3+} ratio). An increased content of ferrous iron (Fe^{2+}) will shift the color to more greenish. In the Netherlands, more than 60% of the recycling glass in the year 2008 is separated at the source (collection banks) or additively in cullet treatment plants. For ultra-white / clear glass, generally only <u>internal</u> cullet is used.

3.3 MOISTURE CONTENT

During transportation and storage of the cullet, the moisture content of the cullet can increase up to about 6 mass%, because of atmospheric conditions (rain, snow). On the average, the moisture content is between 1 and 3%. Finer cullet may contain much more moisture, because the larger surface area will adsorb much more water and will hardly release this adsorbed humidity. A very high moisture content can have an influence on the oxidation state of the glass melt and therefore may harm the melting process (it can create foam formation and fining problems). In addition (too) high moisture content will lead to unnecessary energy losses, because the water has to be vaporized and water vapor is heated to the glass furnace temperatures. Control of the moisture content (for instance by covering the cullet heaps or by dry storage) therefore is important for a stable melting process and to reduce energy consumption. Changes in the amount or concentration of dissolved water in the molten glass will slightly change some properties (viscosity, surface tension, heat transfer), for instance important in the forming process. Water content in the glass, depends hardly on the water content in the batch, but is more strongly influenced by the water vapor content in the glass furnace atmosphere [11].

3.4 CULLET SIZE (-DISTRIBUTION)

The cullet size distribution [12], especially the fine (powder) cullet fraction, has an important effect on the melting process. Usually the fraction of fine cullet (typically < 0.2 - 1 mm) contains more moisture and organic impurities than the larger cullet pieces. The fermentation process of organic materials in fine (polluted) cullet is hindered or limited by the very slow air infiltration in a pile of fine cullet. Fermentation of organic material is faster in coarse cullet because of faster oxidation (oxygen supply) in the interior of cullet heaps.

High contributions of fine cullet in the batch therefore may cause problems in the melting process like foam formation and oxidation state variations. With higher amounts of fine cullet, more carry-over will occur, causing a faster attack of the refractory.

3.5 SEPARATION / SORTING OF CONTAMINANTS FROM RECYCLING CULLET

The methods for cullet sorting and removing of contaminants as described below are often applied in the glass cullet treatment plants. A waste glass cullet purification line consists of combinations of sorting and separation technologies given in the following sections.

Separation of ferro- and non-ferro metals from cullet [13]:
For the separation of ferro-metals, magnets are applied, but it is also possible to use equipment which separates almost all metal pieces larger than 1 mm, in one separation process. A stream of cullet is distributed from a conveyor to a wide drain (gutter) where the glass pieces are nicely distributed in one layer. Prior to the removal section, a detection system is installed, existing of a multi-canal high frequency coil, separated in segments of a few centimeters of which each detects a passage and specifies the location of ferro metal pieces as well as a non-ferro metal in the cullet stream. The detection is based on an interruption of an applied electromagnetic field by ferro- and non-ferro metallic contaminants:

A modern detection unit (a row of such units in series can be installed all over the width of the conveyor belt) for metal pieces exists of a coil, sending a high frequency **electromagnetic** alternating field with a frequency of 10 up to 600 MHz, depending on the application.

By this way, a piece of metal in a well distributed cullet stream can be detected. Electronic devices control the mechanical separation of the metal pieces as they pass by, using pressurized air flowing through a nozzle or by using a mechanical separation system that pushes out the metal piece from the glass cullet flow. Such a system can be integrated in the separation facility of a glass recycling company (CTP), but can also be applied in the glass industry itself.

Other systems, like eddy-current metal separators are used for the separation of ferro and non-ferro metal pieces from cullet. Permanent magnets can produce an alternating magnetic field, for instance by rotating these magnets. Such an alternating magnetic field, generates electrical currents in materials with a sufficiently high electrical conductivity, like most of the metals. A high conductivity causes a high electrical current in the material and a lower conductivity will give smaller currents. The electrical current in ferro and non-ferro metals, passing the alternating magnetic field will result in a Lorentz force (due to the magnetic field and the electrical currents) acting on the metallic pieces.

Due to the different Lorentz forces (and no force acting on the glass pieces), the trajectories of the different metallic pieces will be different from the cullet flow, but depends also on the kind of material. The different materials can be separated by collection of the pieces depending on their trajectory.

Separation of ceramics, stones and china ware from cullet streams
Generally, these kinds of contaminants are non-transparent or only opaque and therefore can be distinguished from transparent glass pieces by optical systems, using lasers. It is possible to separate these kinds of ceramic pieces from glass cullet at sizes down to 2 mm. But the detection and separation efficiency drops at these small sizes. The separation of ceramic pieces, stones or chinaware from glass cullet can be combined with the metal separation process.

The cullet is transported by a conveyor to a wide drain on which the single cullet pieces are distributed. A laser camera with independently working laser diodes constitutes the detection system. The transparent glass pieces transmit the laser beams, but when a non-transparent piece passes the beam, the laser signal will not reach the detector. Each sensor will send a signal a few thousand times per second, giving a very good resolution of the presence and position of ceramic impurities in the cullet stream. Pieces of 2 mm and larger can be detected and can be separated by an air jet which is driven by the signals of the detectors. Different systems for these separation functions are available yet. Separation of glass-ceramics (often transparent) is hardly possible with these devices because glass-ceramics or other glass types are transparent. Glass-ceramics hardly melt or only slowly dissolve in soda lime silica glass melting tanks.

Separation on color [literature reference 14]
Sorting of glass cullet on color, by opto-mechanical devices is being applied in the last decades. First, only coarse glass pieces could be separated on color, especially when several tons of cullet per hour (> 10 tons/hours per machine) have to be treated. Today, systems are available which separate cullet pieces on color even in the size range of 3-5 mm. Especially, for the production of high quality flint glass, the fraction of green glass in the white recycling cullet may not exceed a level of 0.05-0.1 mass-%, because this would lead to extra green colorizing of the glass. Systems for separation of coarse cullet on color are often preferably used in the first stages of the cullet treatment process before the glass has broken too much.

So called ultra-high power laser sources send converging multi-color laser beams to the single glass pieces, and scan in the direction the cullet falls at distances of 2 mm. The transmitted light of the laser beams is detected by optical devices. An exact determination of the color even for polluted glass pieces appears to be feasible. The electronic system determines the location of the single pieces and the time a glass cullet piece of a certain color passes the air jet system. By this way, flint cullet can be separated from colored cullet. Green cullet can be separated from flint and amber, or amber glass can be separated from green and flint glass, depending on the kind of color of cullet, which is required.

For large quantities of cullet to be sorted, systems can be applied which separate the coarse fractions on color. Sorting on color of finer fractions will decrease the amount of glass, which can be treated during certain time period. For high throughputs, after sieving, only the coarse pieces larger than about 8 mm are being separated on color. The finer fractions of mixed cullet (mixed colors) can be used for the production of green glass. Of course, metallic pieces and ceramic materials larger than 1.5 to 2 mm should be separated or the cullet should be grinded to sizes smaller than about 1.5 mm in order to obtain no glass defects after melting. It is recommended to remove almost all metal contamination, even small pieces of metals can cause severe glass quality problems or attack of glass melting tank bottoms.

Sophisticated detectors use filters which only detect a very small selected part (band width about 20 nm) of the spectrum. These systems use convergent white light which is partly transmitted through the passing cullet pieces. Only the part of the spectrum for which the glass piece is transparent will be transmitted. Because different glass colors and even different glass compositions will show different absorption spectra, a combination of detectors, each sensitive for only a small part of the spectrum can identify which kind of color or glass passes the detection area. Ceramic pieces will absorb visible and UV-light almost completely and can be distinguished clearly from the more or less transparent glasses.

The total transmission of a piece of glass depends also on the size (thickness: optical path through a cullet piece), and therefore it is not very easy to separate the different kinds of glasses with different sizes or thicknesses. Because the absorption spectrum (the total absorption, depending on the wavelength and thickness / optical path) depends on the size of the glass pieces, today the cullet is preferably separated on size before the color separation.

Hyper-spectral imaging systems are currently in development to distinguish different types of glass from each other by their different spectral properties, depending on glass composition.

Removal of Glass Ceramics from Recycled Soda-Lime-Silica glass cullet
The mechanical removal (pressurized air pulse or other mechanical device to remove detected impurity from cullet flow) itself can be done with techniques that are also used for colored glass or non-transparent contaminants like CSP (Ceramics, Stone, Porcelain). Critical is the detection of glass-ceramics pieces and their position in streams of cullet. The applied non-contact methods are based on a different behavior of "soda-lime-silica glass" and typical "glass-ceramics" while being exposed to a certain electromagnetic wave source (e.g. visible light, UV, IR, X-rays). The detection is done while the cullet is on a conveyor belt or is in free fall. The composition of glass ceramics is very different from common soda-lime-silica glass types, most glass ceramic materials typically contain large amounts of Al_2O_3: 20-25 mass%, 2-3 % TiO_2, up to 5 % Li_2O, sometimes a few % P_2O_5, 1-3 % ZrO_2, 0 to 2 % ZnO, sometimes As_2O_3 (0-1 %) or tin oxide. Total alkali oxide contents are generally below 5 mass%. SiO_2 concentrations are typically around 65 mass%. The viscosity of this glass, at for instance 1400 °C, is much higher than for soda-lime-silica glass.

Glass-ceramic detection
Glass-ceramics resembles the glass cullet in many aspects; density, shape, size and color(s). Glass-ceramics is in contrast to glass a partly-crystalline material. The controlled formation of crystals during the (slow) cooling of a glass, in a certain temperature range, is obtained by applying nucleation agents in the glass forming raw material batch. Nucleation agents are compounds present in the material in concentrations of a few percent (or less) and are responsible for the onset of the crystallization during thermal treatment of the glass.

Electromagnetic waves of the right frequency will interact with glass-ceramics in a different manner than other glasses, due to the presence of these dopes or the presence of other chemical elements that are not present in normal glass types (e.g. soda-lime-silica glass). In most cases, titanium oxide or zirconia is added as nucleation agent. Due to the presence of these dopes or other chemical elements not being present in standard soda-lime-silica glass, the detection might be done by the analysis of (a part of) the absorption- or luminescence spectra of the different types of glass and glass ceramics. Different absorption and reflection spectra (finger prints) can be found for different glass types. The investigated techniques are discussed in detail below.

Absorption
Due to interaction of electromagnetic waves with the material (glass or glass-ceramics), a part of the exposed energy (light or X-rays) will be absorbed while travelling through the material. This absorption is a function of the wavelength and the transmission / absorption spectrum is characteristic for the composition of the material. The intensity depends on the thickness of the material, but the shape of an absorption spectrum is a fingerprint of the material. The resemblance between glass and glass-ceramics is large, so effort should be spend to find the

range of wavelengths which are most distinguishable. In literature, three ranges of wavelength that were investigated are found; infrared, ultraviolet and X-ray.

- Infrared

The University of Rome (in cooperation with Reiling Recycling Germany) published a study [15] on the identification of glass-ceramics from other glass types by using mid-infrared light. The features of an infrared spectrum are directly related to the molecular structure of a material, hence these wavelengths are investigated. It appeared, that the ratio of the intensity of two narrow wavelength bands (2.85 and 3.65 micron), shows a difference when comparing this ratio measured for normal glass and for glass-ceramics. See figure 7.

At an experimental setup, more than 85% of the glass-ceramic samples could be detected correctly. The typical size of the studied cullet pieces is 5 mm and larger. Since only a few narrow wavelength bands are used, the costs of a detector will be reduced significantly, but at the cost of a lower sensitivity. At the moment, the technique is in a research state.

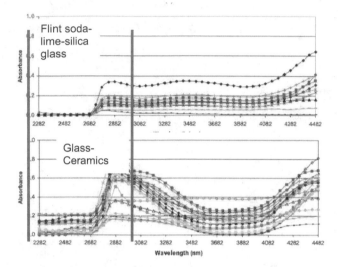

Figure 7   The ratio of measured absorption at 2850 and 3650 nm is used to discriminate SLS glass (upper graph) from glass-ceramics (lower) [15].

- Ultraviolet

Another option is a sorting machine for glass-ceramics based on transmission or absorption characteristics, mainly in the ultraviolet region and partly in the visible region of the spectrum. The technique is based on material specific UV-absorption edges and "UV-cut off frequencies". Reference values for UV-absorption and "UV-cut off frequencies" from standardized materials (soda-lime-silica glass and "glass-ceramics") were determined by laboratory analysis first. It seems that the measured reference values differ at a level that soda-lime-silica glass can be discriminated from "glass-ceramics". This makes the "cut-off" a clear criterion for differentiation. To grant highest efficiency and achieve lowest possible amount of false detection, the colour information in the visible wavelength is analyzed additionally. According to the supplier, the efficiency of the machines is 86% and can detect glass-ceramics larger than 4 mm. The cullet sorting capacity is 3-15 tons/hour and is currently used by several cullet treatment plants in Europe. Cullet sorted by these sorting machines appeared to reduce drastically the machine jams that were often caused by glass-ceramics at glass plants of container glass producers in Europe.

- X-rays
Another option is the application of X-rays. A machine with an X-ray source plus detector is applied in this method. A conveyor belt with cullet on it moves between the X-ray source and detector device. The fingerprint of all the known glass-ceramic samples is compared with the transmission spectrum of each glass piece passing.

In performance tests, less than 2% of glass was wrongfully detected as glass-ceramics and when the cullet size is larger than 10 mm, over 90% of the present glass-ceramics has been removed correctly. In case the size of glass-ceramics is smaller than 10 mm, the detection efficiency appeared to be drastically lower (around 40-50% for particles of 6-8 mm). Some typical specifications are listed below:

- Sample size cullet pieces: 10-60 mm
- Capacity per sorting machine: max 20 tons cullet/hour
- Belt speed: up to 2.5 m/s
- Requirement: Single glass cullet layer on conveyor belt

These types of sorting equipment are supplied by few equipment companies.

- Luminescence
The absorption of the high energetic X-rays results in the transmittance of fluorescence light with specific wavelengths, depending on the chemical elements in the glass. The method that analyses this fluorescent light spectrum is called X-ray fluorescence (abbreviated to XRF) and is widely applied as a detection technique and the principle is since long time used for chemical analysis (chemical elements in materials). Even small handheld detectors are available to check the type of glass. Sorting machines for the glass recycling industry based on XRF are for instance produced by the American company Innov-X. They claim to have a detection yield of 95% at a rate of 20 tons cullet /hour (it is not specified for which sizes of glass-ceramics this detection yield is valid).

Hyper-spectral imaging
An intense light source is used to illuminate glass cullet moving underneath the light source. The transmitted or reflected light from different areas is received by spectrophotometers. The spectra have a high resolution and the received spectra are compared with spectra of different glass and glass ceramic types. Software is developed to distinguish the different spectra and relate these to different glass compositions. This detection method is in development for application of detecting different types of glass, such as glass ceramics and lead crystal glass in a stream of soda-lime-silica glass cullet.

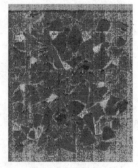

Figure 8   Glass (soda-lime-silica) cullet with some glass ceramics pieces
At the left hand side the normal image, at the right hand side the process image using hyper-spectral imaging for the same layer of cullet. The positions of glass ceramics (yellow and red) can clearly be observed from the right hand sided image.

This detection technique, in combination with devices to remove the detected glass ceramic cullet pieces, presents a promising method to sort out glass ceramic pieces of 4 mm size and larger.

Hyperspectral Imaging (or also called Imaging Spectroscopy) is a technique that uses a combination of Imaging and Spectroscopy:

Imaging: Measurement of light (radiance) as a function of spatial position
Spectroscopy: Measurement of light (radiance) as a function of wavelength

Different substances have a different way in which they reflect or absorb radiation. This difference in reflectance or absorbance (as function of the wavelength) enables a possibility for an automatic image processing and to discriminate between several types of substances (e.g between soda-lime-silica and glass-ceramics).

Figure 8 shows the result of a composed image after processing the spectral data and converting it into a new image with detected glass ceramic pieces in a layer of soda-lime-silica glass cullet.

Checking the cullet quality
External glass cullet (post-consumer) is one of the main raw materials in glass production for several glass types. Especially recycling cullet needs to be examined regularly on its composition and on its contamination levels by different possible impurities.

The glass industry (container glass, float glass, glass wool, domestic glass, specialty glass) needs a very strict sampling and checking procedure for the cullet quality before the cullet can be recycled in glass furnaces, because of the ever increasing ratios of cullet in the batch and the effects of contaminants like metallic pieces (steel, aluminum, nickel etc.), ceramics, stones, chinaware (also glass-ceramics) on the glass quality (stones, silicon inclusions, NiS-inclusions) or melting process or even glass product forming process. Metal pieces in cullet may lead to tank refractory damages, such as downward drilling due to metallic material sinking to the bottom of a glass melting tank.

Manually sorting of a representative sample of cullet on metals, stones, ceramics, organic materials, plastics, different glass colors can be applied as well. The main difficulty is the detection of glass ceramics or lead crystal glass pieces in the cullet. Sampling of the cullet is an important element in this procedure. Generally, cullet samples of a few up to 16 kg from different areas of large cullet piles (lots of 500 to 1500 tons glass cullet) are collected and combined and mixed together. After mixing, typically a sample of 50 kg of this cullet is taken for more detailed analysis on impurities.

Instruments or equipment has been developed to check large cullet samples (10-100 kg) or a continuous flow of cullet on the quality. These instruments can be used at the cullet treatment plant (CTP) to specify the cullet to be delivered to the glass factory. Some glass companies, first test the cullet quality before the cullet is shipped from the CTP to their factories.

In such installations, larger quantities of cullet are automatically transported to a storage container and continuously are being sieved in a sieving drum, separating the fractions in the ranges 0-8 mm, 8 to 16 mm and 16 up to 60 mm. With optical detection systems, the contents of ceramic pieces, stones or china and the amount of metallic contaminants are measured and can be continuously checked for each size class.

Special aspects for recycling in float glass and insulation glass wool or fibers
In the float glass industry, about 20 to 30 % of the batch consists of own cullet or cullet from neighboring glass users. The legislations in Europe, especially in Germany, force the glass industry to recycle more and more waste glass from post-consumer automobiles and buildings.

But, due to the high quality demands of the float glass products (especially for automotive applications or mirrors) on one hand and on the other hand the presence of other material types sticking to the waste glass (cements, ceramics, organic substances, enamels on automotive glass), the recycling of this glass in float glass furnaces is hardly possible.

Future developments in the separation of polymers (for instance for laminated glasses or the sealants in double-glazing units) from the glass, will contribute to future potential for more and more recycling of waste flat glass.

Extremely important is the elimination of all metallic components/impurities especially nickel or nickel containing steels (e.g. stainless steels) and glass ceramic pieces > about 2-3 mm, from the cullet to be recycled. This is

extremely important for using recycling glass in flat (float) glass furnaces. Nickel containing steel flakes can lead to formation of nickel sulfide inclusions in the glass.
During tempering of the glass (e.g. thermal toughening) this nickel sulfide can undergo phase transitions. The very small nickel sulfide inclusions can cause spontaneous collapse of tempered glass panes containing these inclusions, even after several years of application (e.g. in buildings).

In the glass wool industry and glass fiber industry, recycling of these glass products [14] will lead to chemical reduction of the glass melt by the organic coatings or binders. This will lead to foam formation, problems in the removal of the gaseous inclusions (fining), change in the color and often a change in the mechanical strength (extremely important for continuous E-glass fibers). It seems that reinforcement glass fiber production needs a rather oxidized glass to avoid high levels of fibre breakage during the drawing process. Therefore it is important to control the oxidation state of the glass melt (and incoming batch, including recycled scrap) in glass fibre production.

Waste glass wool and glass fiber products can be cut into small pieces and can then be treated thermally in rotary kilns in order to remove the binders or coatings, partly by burning or volatilization of the organic components. The fibers are clean after the thermal process and can easily be used as a raw material for glass wool, but also for E-glass production. The gases released in the rotary kiln have to be treated in a post combustion chamber, the exhaust gases of this chamber are partly released or are used to heat the thermal treatment of the waste fiberglass to be recycled.

3.6 QUALITY DEMANDS RECYCLING CULLET FOR GLASS INDUSTRY

The FEVE (Fédération Européen du Verre d' Emballage: European federation of container glass manufacturers) provides specifications for processed recycling cullet with respect to the quality aspects as described before. Glass companies also define their own specifications. Table 2 shows a typical specification list for recycling cullet being suitable for container glass production.
To check the cullet quality, typically a representative cullet sample of a size of 0.05 % of the total lot (total charge of cullet) has to be analyzed. For one truck load about 30 - 50 kg cullet has to be sampled to check the specification criteria.

Sometimes, automatic sampling of cullet from the cullet treatment plants is applied. A representative sample of 15 kg is analyzed on humidity, ceramics, stones, cork, paper, plastics, metals etcetera. Especially the analysis on ceramics, stones, china (porcelain), opal glass, magnetic metals and non-magnetic metal, including lead is done very regularly (on about 1 % of the cullet to be delivered).

In high quality flint glass cullet, the amber glass content might not exceed 0.2 mass-% and the green glass content should be lower than 0.05 %, but for some flint glass qualities up to 7 kg green cullet can be accepted per ton recycling glass. In amber cullet, the fraction white should preferably be limited to 8-10 % and the fraction green should not exceed the 8-10 % level.

In sorted green glass cullet, up to 15 % flint and up to 5 % amber glass can be accepted. For amber glass production, relatively large quantities of float glass cullet can be used in combination with reducing agents to convert the sulfate into sulfide.

Table 2 Some specifications for processed recycling cullet to be applied in container glass furnaces, source FEVE and others. Different specifications can be applied for these cullet qualities by different glass factories.

Cullet specifications	acceptable cullet (units in mass % or grams/ton cullet)
stones, ceramics, chinaware, pottery excluding glass ceramics	< 25-35 g/ton
glass ceramics	indicative < 25 g/ton
glass ceramic pieces	if present, size should be < 3-4 mm
magnetic metals	< 5 g/ton
non-magnetic metals	< 5 g/ton
lead	< 1 g/ton
aluminum	< 5 g/ton
all metals	< 7 g/ton
organic material	< 200 or 500 g/ton
COD of washing water from cullet	< 1200 -1500 mg O_2/liter
plastics	< 60 g/ton
moisture	< 2-3% (preferred)
paper/cork/wood	< 1500 g/ton
opal glass	< 100 g/ton
grain size cullet	no cullet pieces > 7 cm cullet pieces < 0.5 cm: max. 12%

3.7 CONTROL OF REDOX/OXIDATION STATE OF RECYCLING GLASS

When using large quantities of external (post-consumer) recycling glass in the raw material batch, for instance in percentages up to 60-75%, the control of the oxidation state (also called "redox" state) of the recycling cullet is very important for a stable melting process and to obtain the required glass color [17-19]. In an oxidized glass or glass melt, oxygen gas is dissolved at a relatively high concentration level. The oxygen vapor pressure in equilibrium with the dissolved oxygen gas in the melt has a relatively high value in an oxidized glass melt. This oxygen pressure value is called the **"pO_2"** of the glass melt (this value is temperature dependent). In oxidized glass

melts, the polyvalent elements are mainly present in their most oxidized valence state (e.g. Fe as Fe^{3+}, sulfur as S^{6+} or sulfate, antimony as Sb^{5+}). In glass melts obtained from raw material batches or recycling glass cullet with reducing agents (organic components, cokes), the oxygen equilibrium pressure is low and the polyvalent ions are mainly in their reduced forms, some metal ions may even be reduced to their metallic forms (lead, copper). The oxidation state here is defined as the ratio between the oxidizing and reducing components in the glass. The oxidation state of glass or glass melt strongly affects both the melting process (fining process, foam formation, energy consumption, SO_2-emissions) and the product quality (color, occurrence of (amber) cords, presence of seeds).

For instance organic materials in the raw material batch and recycled cullet may lead to the following reactions during batch blanket heating and melting-in, in a glass furnace:

During heating in batch blanket:
- Pyrolysis: CxHyOz ⇨ cokes/char, CO, water vapor   1]
- Boudouard reaction: 650-1000 °C:
    C(char) + CO_2 (from decomposed carbonates) ⇨ 2CO   [2]

In temperature range: 650-1100 °C:

650-850 °C:

Sulfide formation $\quad\quad 4CO + SO_4^{2-} \quad \Rightarrow S^{2-} + 4CO_2$ [3]

900-1100 °C:

Sulfate decomposition: $\quad CO + SO_4^{2-} \quad \Rightarrow SO_2 + CO_2 + O^{2-}$ [4]

Sulfide-sulfate reactions after batch melting (1100-1350 °C):

$S^{2-} + 3SO_4^{2-} \quad \Rightarrow 4SO_2 + 4O^{2-}$ [5]

causing primary foam

Ferric iron reduction may occur during batch heating at reducing conditions:

$$CO + O^{2-} + 2Fe^{3+} \Rightarrow 2Fe^{2+} + CO_2 \quad\quad [6]$$

The direct or indirect chemical reduction by organic materials present in the cullet will give the cullet **"reducing" power**. Removal of organic contaminants will decrease this reducing power.

When using cullet with high reducing power, SO_2 gas evolution at low temperatures (< ±1250 °C) will cause foaming (equation 5). The valence state of the polyvalent ions in the melt will be reduced by the reducing agents and this changes sulfur behavior, glass color and radiant heat transfer.

The "redox" state or reducing power of the recycling glass in the glass melting process is determined by the following factors:

- Color composition of the cullet (percentage float, white flint, green and amber glass or other reduced glass types);
- Type and content of organic contaminants (paper, sugars, fats, plastics, etc.): sugar in the batch has much more reducing power (1 kg sugar can be as reducing as 20 kg paper) than paper;
- Cullet size (distribution) combined with moisture content and organic components in cullet.

A fluctuation of the quantity or type of organic material in the recycling cullet can be caused by a change of the cullet source, change of cullet storage time or change of outside temperature. The temperature and storage time will have an effect on the breakdown of organic components in the contaminated cullet by fermentation processes.

Natural fermentation processes (by micro-organisms in the cullet pile) will decrease the amount of organic components and will lower the reducing power of the cullet. Fermentation processes in piles of cullet powder are often very slow, because the air can hardly infiltrate in the inside sections of this cullet heap. Therefore, fine cullet often shows a higher reducing power compared to coarse cullet from the same sources. Fermentation depends on temperature, humidity and oxygen level in the cullet pile.

Changes in oxidation state of the cullet in the batch will cause changes in the Fe^{2+}/Fe^{3+} concentration ratio in the obtained glass melt and glass product and causes changes of the sulfate or sulfide retention. These properties will influence the glass color. The dominant wavelength and the color purity can fluctuate when the redox state frequently changes.

Several methods have been developed to characterize the redox state of recycling glass cullet, namely:

1. COD method for the washing water, after washing the cullet with hot water;

2. Calculation of redox number of cullet or batch from redox contributions of the different glass colors and known quantities of organic contaminants in the cullet;

3. pO_2 measurement of a cullet melt;

4. Measurement of mass loss after heating pre-dried cullet to a temperature of 900 or 1000 °C, at which all organic materials are burnt off;

5. Melting of the cullet, after melting cooling this melt, annealing the glass and measuring the glass transmission spectrum. The absorption of light at 1050-1060 nm (caused by presence of ferrous iron in glass) in relation with the total iron content in the glass will be a measure for the oxidation state of the glass obtained from this cullet.

The first three methods are explained in more detail:

1. Determination of the **chemical oxygen demand** (COD) of the water-soluble part of the organic components in the recycling cullet.

In practice this method is applied in a number of situations where color composition of the delivered cullet is rather constant, but it has the following disadvantages:

- the non-soluble components (like most plastics) are neglected in the analysis;
- the reducing effect of some organic substances may be different at the high temperatures in the glass melt than at room temperature in a solution used for the COD determination;
- the redox-effect of the different glass colors (for instance the reducing effect of amber cullet) is not measured by this way of COD-analysis.

In summary the COD analysis works as follows: **C.O.D. analysis (C.O.D. = Chemical Oxygen Demand)**

o Washing of a sample of 10 kg waste glass cullet with 65 °C water;
o Soluble organic material will dissolve in this water (aqueous solution);
o The Chemical Oxygen Demand will be determined by the so-called Dr. Lange method, result: mg O_2 demand per liter solution. This method is based on a redox titration of the obtained aqueous solution.

In many specifications, the requested COD level should be lower than 1200-1500 mg oxygen per liter of obtained water after cullet washing. By controlling the COD of the applied cullet, the transmission spectrum and colour of the glass could be stabilized in a glass factory, see figure 9.

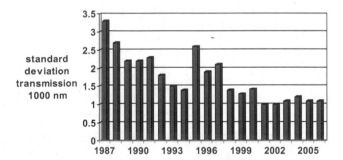

Figure 9   Reduction of the variation in redox state of glass (transmission at 1000 nm) by improved control of recycling cullet COD (Chemical Oxygen Demand)

Figure 10 shows that a chemical reduction of the glass melt by more reduced cullet (high COD levels) may lead to a drop of the temperatures in the bottom of a melting tank. The heat radiation from the flames is absorbed by the increased levels of ferrous iron (Fe^{2+}) in the melt. The transmission of the produced at 900 nm will decrease, because of increasing Fe^{2+} contents in the glass and tank bottom temperatures will decrease at the same time.

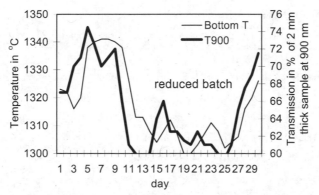

Figure 10 Transmission of produced glass at 900 nm (as a measure of ferrous iron content) and tank bottom temperature during 1 month

2. Calculation of the redox number of the entire batch, including the recycling glass, in which a redox number is attributed to the recycling glass. The redox number of the recycling glass is calculated based on the color composition of the cullet and an average contribution of the organic contaminants [17,18]. Table 3 shows typical redox states of glasses dependent on their color.

In practice, this method also is applied with success. A drawback of this method is that the redox-factors of the different organic ingredients are often not known. Moreover, the quantities of the different organic components may vary widely from charge to charge.

Table 3 Typical redox states, expressed in ferrous iron total iron concentration ratio, for different soda-lime-silica glass colors

Concentration ratio Fe^{2+}/Fe_{total}	Glass colour
0.10 – 0.40	White/slightly yellow
0.20 – 0.60	Green
0.60 – 0.75	Olive green ("feuille morte")
0.75 – 0.90	Amber (brown)

The temperature and process dependent oxidation state of a glass melt can not unambiguously be described by one single variable derived from batch composition. The final oxidation state of the glass or glass melt will not only be determined by the raw materials and their contaminants but also by the melting rate, furnace atmosphere, the direct contact between the batch blanket or the glass melt surface with the flames.

3. Measurement of the **oxygen activity** (this is the equilibrium oxygen partial pressure of the oxygen dissolved in the melt: pO_2) in a test melt of recycling glass at simulated process conditions (simulating temperature, heating rate). A two-electrode system is applied.

The pressure of oxygen in equilibrium with the oxygen dissolved in the melt is determined by dipping in the melt, a zirconia/platina sensor (2 electrode system) in order to characterize the

redox state of the molten glass at a certain temperature [18, 19]. A reference electrode, using clean air (pO_2 = 0.21 bar) or a metal/metal oxide couple as reference state in the reference electrode is used. Between the reference electrode (in equilibrium with air or the metal / metal oxide couple) and the platinum electrode directly dipped in the melt (special design to avoid air infiltration from outside along the platinum wire) the electromotive force (EMF) can be measured in volts or millivolts.

The ion-conductivity between the melt and the reference electrode is provided by a small zirconia holder, also dipped in the melt, the reference electrode is fixed inside the zirconia holder. Figure 11 shows the principle of this method.

The measured EMF-value is directly related to the ratio between the oxygen equilibrium pressure of the melt ($pO_{2\,melt}$) and the oxygen pressure at the reference electrode (pO_{2ref}), by Nernst law. The EMF value is proportional to the logarithmic value of the ratio: $pO_{2\,melt}/pO_{2ref}$. When using clean air in the reference electrode, pO_{2ref} = 0.21 bar. When using metal / metal oxide as reference, the pO_{2ref} is temperature dependent and can be derived from the temperature and thermodynamic data of the pure metal and pure metal oxide.

Different electrochemical sensors have been developed, based on electrochemistry for measuring the oxidation state of the molten glass in terms of the pO_2-value (in bar or Pa). Figure 12 shows an example.

The method for characterizing the oxidation state or reducing power of the delivered cullet using this approach, thus by the principle of the "pO_2"-value is as follows:

1. Representative sampling from the recycling cullet (selecting an average, representative sample from the whole pile). For a batch of hundreds of tons cullet, typically 25 samples of 16 kg cullet from different locations of the cullet pile are taken at the surfaces of the piles plus 25 samples of 16 kg from the inside. These 2 x 25 x 16 kg samples are combined and mixed. From the 800 kg of cullet, after mixing, smaller samples are taken, combined mixed and so forth. For instance from the 800 kg sample, one representative quarter is taken (200) kg, from this sample, again 1 quarter (50 kg) is taken from that part of 200 kg, until the required cullet sample size has been gathered. For an analysis of the oxidation state of the melt by the electrochemical method, about 600-800 grams of cullet are melted.

2. Rapid melting procedure of about 0.6-0.8 kg recycling glass at controlled conditions, followed by pO_2-measurement (using a oxygen sensor: disposable or re-usable) in the test melt at one (e.g. 1300°C or 1400 °C) or more selected temperatures. Figure 12 shows the sensor lay-out for one of the commercially available systems.

3. Prediction of the oxidation state (e.g. Fe^{2+}/Fe^{3+} concentration ratio), the color and the residual sulfate-content of the container glass, produced from raw materials and this cullet, by a computer program, taking into account the thermodynamics of polyvalent ions and oxygen in the molten glass.

With this pO_2 method, also the effects of different measures to control the redox state of recycling glass can be quantified and applied in a controlled way. Potential measures to influence or compensate for the redox state (reducing power) of recycling cullet are:

- "Fermentation" of the organic components in the recycling glass during a number of weeks 5-10 weeks depending on temperature.
  Generally, the quantity of organic components will decrease during these weeks by a factor 2 till 4. Fermentation only takes place at sufficient high ambient temperatures. In wintertime, however, fermentation may be limited or slow, due to the colder temperature levels.
  The fermentation process is exothermic and inside a pile of cullet with organic contamination (fats, sugars, food residues, etcetera), the temperature increases during the start of the fermentation process. At the end, when most organic material is converted (generation of water, CO_2/CO, odors) temperature in the pile decreases again.
- Preheating or rinsing the recycling cullet to remove the organic components.

- Extra addition of oxidants to the batch, like sodium sulfate (gives increased SO_2 emissions and may lead to foaming) or nitrates (increases NOx emissions).

Especially pulverized cullet is very sensitive for the presence of organic components. During the first sintering and melting stages, the gases released during the reactions cannot be freely released because of the sintering process. Reducing gases (CO, hydrocarbons) remain between the sintering cullet in the batch blanket and will chemically reduce the glass components, like sulfate into sulfide or ferric iron (Fe^{3+}) into ferrous iron (Fe^{2+}) during the heating and melting process. The melt shows a low pO_2 value.

Figure 13, presents an example of the measured pO_2 (oxygen partial pressure in equilibrium with dissolved oxygen in glass melt) as function of the temperature of melted 100% (very reducing) pulverized recycling cullet with addition of different amounts of sodium sulfate (see inset). Without sodium sulfate addition, a melt of this fine cullet will show very low pO_2 values (highly reduced melt, with feuille morte glass color). This figure shows evidently that the melt of the pulverized cullet can be oxidized by the addition of sodium sulfate, this increases the pO_2 level (up to 2 orders of magnitude).

During storage of powdered cullet with organic components in the cullet pile, fermentation is slow and the amount of organic contamination can be rather high even after longer storage times.

Figure 13 shows that the pO_2 value not only depends on the cullet or batch composition, but also on temperature. Thus, to characterize the oxidation state of cullet by the cullet melt pO_2 value, the temperature of the measurement should be fixed.

Using, more reduced cullet for green glass production will shift the glass color to amber as figure 14 indicates.

Measuring principle & Set-up

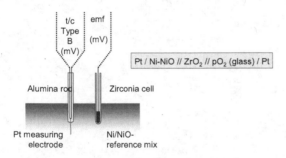

Figure 11   Measuring principle of pO_2 in molten glass by measuring sensor and Nickel / Nickel oxide reference electrode (info from company ReadOX). The oxygen pressure (bar) in the melt pO_2 (glass melt) can be derived from the EMF (in volts) measurement

$$EMF\,(V) = \frac{RT}{nF} \cdot \ln \frac{pO_2\,(\text{glass melt})}{pO_2\,(\text{ref. Ni/NiO})}$$

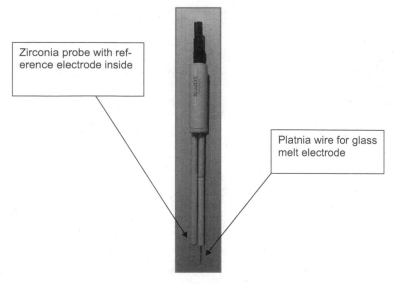

Figure 12 Electrode assembly to measure oxygen equilibrium pressure (pO₂) of the glass melt (info from company ReadOX).

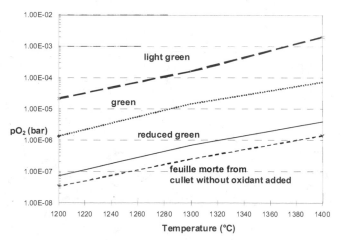

Figure 13 Measured oxygen activity (pO₂) as function of the temperature in trial melts of organic material contaminated pulverized recycling cullet with additions of different amounts of sodium sulfate.

Redox sensors in combination with redox model
Colour prediction

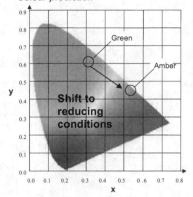

Figure 14   Shift of glass color when glass melt becomes more reducing.

CONCLUSIONS

Waste glass recycling requires strict specifications concerning the maximum levels of impurities that can still be tolerated in waste glass to be recycled in glass furnaces. Cullet treatment plants need to improve sorting performance to remove metallic pieces, ceramics, stones, glass ceramics and to stabilize the contents of organic components in the cullet. Pollution of cullet leads to glass defects, color instabilities or causes process disturbances (foaming, changes in heat transfer, downward drilling) in glass production. Today, effective systems exist to remove ferro-metals and non-ferro-metals, to sort larger pieces of glass on color and to remove ceramics and stones. However, fresh post consumer container glass cullet still contains a high level of organic material that will chemically reduce the polyvalent ions in the melt, resulting in changes in heat transmission in the glass melt tank and color changes: dominant wavelength, color points. Control of cullet redox state and specifications for the COD or pO_2 of the cullet are important for stable operation. Cullet storage during several weeks or washing of cullet or cullet preheating will lower the COD values and reducing power of the cullet to be recycled.

The methods for separation / removal of glass ceramics from soda-lime-silica glass cullet are still under development. Existing equipment can effectively sort glass ceramic pieces > 10 mm but pieces > 3 mm may already be harmful for the glass melt process and product quality (glass ceramic inclusions, tension spots). Therefore new methods for high performance removal of glass ceramics from cullet at a high rate have to be developed.

Cullet is a raw material and it does not make sense to apply very strict specifications to all normal raw materials, but being more reluctant in cullet specs. With high quality cullet, furnaces may operate with up to 95 % cullet with low energy consumption and a low CO_2 footprint. Glass can be recycled endless, but the cullet should be clean.

For green glass production, furnaces in the Netherlands and Germany typically apply 85-90 % cullet in the batch, typically 35 % can be mixed colored cullet and also flat glass cullet is suitable (typically 0-15 %), internal cullet typically 5-10% and color sorted green glass 35%. For flint glass production only color sorted (flint) cullet can be used. Up to 65 % cullet is applied in flint glass raw material batches, including 5 – 10 % internal cullet, and depending on the color specifications flat glass cullet (0-10 %) plus color sorted post consumer cullet. In amber glass production up to 45 % mixed colored cullet is often used plus 5-10 % internal amber cullet and color sorted external amber cullet. In amber glass production the glass is typically produced from about 80 % cullet and only 20% normal batch. The cullet for amber glass production can contain flat glass up to 20 %.

LITERATURE REFERENCES

1. FEVE website: http://www.feve.org/
2. A.J. Faber; R.G.C. Beerkens; C.Q.M. Enneking: *Recycling van verpakkingsglas*, Klei Glas en Keramiek No. 9 (1992), pp. 276 – 279
3. C.Q.M. Enneking; R.Beerkens and A.J.Faber: *Glass melting from cullet rich batches*, Glass Industry, July 1992, 18 – 21
4. H.A. Schaeffer: *Recycling of cullet and filter dust - Challenges to the melting of glass*, Proc. Int. Symp. Glass Sci. And Techn. Oct. 1993, Athens, p 21-29
5. Integrated Pollution Prevention and Control Reference Document on *Best Available Techniques for the Manufacture of Large Volume Inorganic Chemicals* – Solids and Others Industry. August 2007 Chapter 2: Soda Ash pp. 32, table 2.1
6. G. Lübisch; W. Trier: *Energiebedarf bei der Herstellung von Behälterglas in Abhängigkeit vom Scherbenanteil*, Glastechn. Ber. 52 (1979) Nr. 6, pp. 141 – 142
7. W.Trier: *Zum Energiebedarf bei der Herstellung van Glasbehältern*, Glastechn. Ber. 55 (1982) nr. 6, pp. 130 – 134
8. P. Buchmayer: *Anforderungen an die Rohstoffe für die Hohlglasindustrie*, in HVG Fortbildungskurs 1988: Rohstoffe für die Glasindustrie
9. M. Beutinger: *Einsatz von Recyclingglas in der Hohlglasschmelze*, Glastechn. Ber. Glass Sci. Techn. **68** (1995), Nr. 4, N51 - N58
10. W.L. Dalmijn; J.A. van Houwelingen: *Glass Recycling in the Netherlands*, Glass, April 1996, 137 – 141
11. P. Laimböck: *Foaming of Glass Melts*. PhD-thesis Eindhoven University of Technology (1998)
12. G. Pirker; A. Gleisdorf: *Optimale Altglasaufbereitung zwischen 4 und 15 mm*. Glas Ingenieur **2**, 1997 pp. 56-57
13. S+S Metallsuchgeräte und Recycling GmbH, Schönberg, *Glasrecycling mit System*, Glas Ingenieur **2**, 1997, pp. 28-32
14. R. Fechner: *Sortieren von Altglas nach Farben*, Glas Ingenieur 2, 1997, 58-60
15. G. Bonifazi; S. Serranti: *Imaging Spectroscopy based Strategies for Ceramic Glass Contaminations Removal in Glass Recycling*. Waste Management (2006) **26** pp. 627-639
16. H.Drescher: *Neue Wege bei der Aufbereitung von Glasfaserabfällen*, Glas Ingenieur, **2**, 1997 pp. 52-55
17. M.Nix and H.P. Williams: *Calculation of the redox number of glass batches containing recycled cullet*, Glastechn. Ber. **63K** (1990), pp. 271 – 279
18. R.G.C. Beerkens; A.J. Faber; J. Plessers; T. Tonthat: *Measuring the Redox State of Cullet*. Glass, October 1997, p. 372
19. A.J. Faber et. al.: *Redox control of recycled glass cullet*, Proc. 4th Int. Conf. Adv. In Fusion and Proc. of Glass, Würzburg, May 22 - 24, 1995, Glastechn. Ber. Glass Sci. and Techn. **68 C2** (1995), pp. 270 – 277

CHARACTERIZATION AND IMPROVEMENT OF GOB DELIVERY SYSTEMS

Braden McDermott, Xu Ding, Jonathan Simon
Emhart Glass Research Center
Windsor, CT

INTRODUCTION

In an individual section (IS) glass machine the gob delivery system consists of three major components that are used to transport the gob from where it is cut to its final destination in the blank molds, Figure 1 is a depiction of this. The scoop is the first component that comes into contact with the gob; the scoop is used to change the direction of the gob from vertical to the correct angle for the trough. Next, the gob will travel through the trough to the deflector. Troughs come in a wide variety of lengths, and materials. Transportation along the trough should not excessively elongate or slow down the gob. After the trough the gob goes through the deflector and is forced to make a directional change once more. The deflector accepts the gob from the trough at the angle of the trough, from here the gob must be redirected into a vertical drop in order to correctly load into the blank molds. In total, the gob must change directions twice and transition out of and into three separate components. Each of these components can lead to some form of gob variation.

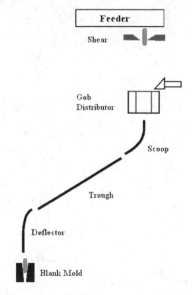

Figure 1: Gob delivery schematic for an IS machine

Variations in gob speed, length, and relative arrival time can lead to improper loading as well as final bottle quality defects. Usually these problems begin as something small but grow larger and larger over time eventually needing to be addressed because they are causing some sort of delivery related issue. The nature of these problems makes it difficult to detect when a problem is starting to occur and what component it is occurring on. To learn more about the variations that occur, due to the delivery system, it became important to instrument the delivery system in a way that gob speed and length data could be collected over a set period of time.

GOB MONITORING SYSTEM

A gob load monitoring system concept was developed to provide real- time blank mold loading monitoring and analysis information. The monitoring system is capable of alerting operators of an impending problem. Utilizing this approach it will be possible to prevent a large loading problem by addressing the issue as soon as the monitoring and analysis system indicates a potential problem.

The monitoring system utilizes fiber optic light sensors, mounted at the shears prior to the scoop, trough entry, trough exit, and deflector exit. Each sensor is enclosed in a robust housing which provides cooling and mechanical protection, to ensure that they are suited to the environment on the IS machine. Each sensor head consists of two fiber optic light sensors separated by a known distance. As the gob passes the fiber optic the light from the gob is captured and converted to a digital signal. With a known distance between the two sensors it becomes possible to calculate the length and velocity of the gob as it passes the sensor head. Figure 2 shows views of the CAD model and the actual sensor heads prepared for a Triple Gob IS section at the deflector exit.

Figure 2: Gob monitoring sensor heads

The length and the velocity calculated using the sensors was verified using high speed photography. For length measurements the uncertainty of the sensor is 5 [mm] and for velocity measurements the accuracy is 0.1 [m/s]. Using these sensors, gob length and velocity were recorded for one section of the IS machine. Over a period of a few months an exceptional amount of data was collected in order to look into the variations that are caused by the gob delivery system and to see if any new information could be learned from the instrumentation of the machine.

STUDY ON GOB VARIATIONS

A series of experimental studies leads us to conclude that gob shape variations occurring before the gob enters the deflector, will be carried on to the blank mold. Contrary to conventional wisdom, the gob shape does not recover as it goes through the deflector, any variations imparted upon the gob will be carried to the blank mold. The study also shows that the misalignment among the gob delivery path components, as well as, the impact of the gob transitioning between delivery components, will introduce a significant variation in gob velocity, length, and arrival times at the blank mold. Any severe variations mentioned above may result in final bottle quality inconsistency. Consequently, gob delivery variations may result in a poor gob load at the blank molds. This could potentially jam the section and lead to machine down time.

Figure 3 represents a typical 15 days of gob speed measurements on a glass machine. From the 9th to 11th a new coating was going through a break-in period. After the 11th the gob speed was stable at the shears, trough, and deflector. On the16[th], gob speed showed more variation at the deflector exit. The engineer inspected the coating on the trough and deflector. It was found that the epoxy coating had worn off at the deflector landing zone. From the 26th to the 30th, the gob speed varied in both the trough and deflector measurements. It was found that the epoxy coating was worn off at the trough entry point. After swabbing the trough entry, gob speed became stable as seen for the final data in Figure 3.

The test results indicated that the epoxy coating had worn off first in the gob impact regions of the deflector and trough. Once bare metal is exposed in the gob impact area, large gob speed and length variations are seen. Adding lubrication in the impact area helps reduce gob speed variations.

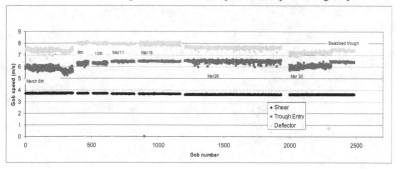

Figure 3: Gob speed measurements on glass machine

Based upon the knowledge gained from the gob variation study, a new design approach was taken for the delivery system. A gob stabilizer was designed to assist the gob when dropping into the scoop, with the stabilizer the gob is guaranteed to drop into the scoop vertical and centered. Bezier curve generation, normal force analysis, and CFD simulations have been applied to new scoop and deflector designs to optimize the gob shape and reduce gob impact along the delivery path. In addition, a new design concept is integrated with the trough and deflector to introduce an air cushion support. This leads to a lower impact and less friction on the gob. Prototype tests showed good improvements in gob shape and reduced gob delivery variations

GOB STABILIZER

There are two main reasons for gob variations: gob misalignment within the delivery system, and impacts caused during the transition of the gob between the components of the gob delivery system.

The first Gob misalignment occurs as the gob loads into the gob distributor head. Emhart Glass developed a gob stabilizing unit mounted directly in the funnel area of the gob distributor. These units have the effect of straightening and centering the gob to help in reducing variations as the gob passes through the scoop.

The series of images in Figure 4 shows the gob positioning immediately after shearing. The top image depicts the situation without the use of the stabilizer - all 3 gobs show a noticeable deviation from the vertical, due to the gob shearing. The lower images show the gob positions after a stabilizer has then been installed for test purposes at the center cavity only. Each image represents the situation resulting from increasing the stabilizing air pressure, demonstrating the verticality improvements for the middle gob. The black bar cutting through the middle gob is from a water hose that was in the way during high speed video acquisition and can be ignored.

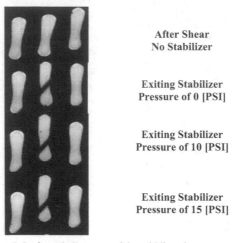

**After Shear
No Stabilizer**

**Exiting Stabilizer
Pressure of 0 [PSI]**

**Exiting Stabilizer
Pressure of 10 [PSI]**

**Exiting Stabilizer
Pressure of 15 [PSI]**

Figure 4: Improved alignment of the middle gob

SCOOP DESIGN

An approach to generating different scoop trajectories, while maintaining curvature has been developed using Bezier curves to represent the scoop trajectory. Bezier parameterization produces generally well behaved scoop profiles. Using this approach, discontinuities and reversals in curvature can be avoided.

The normal load analysis approach was developed to ensure a smooth and consistent normal force pattern and minimized peak normal load applied on the passing gobs, while maintaining a smooth increase and decrease in the normal force. With gradual increases and decreases in the normal force the scoop should have less of an impact on the gob shape. With a minimized normal force, less friction should be expected between the gob and scoop, which means a minimal influence on gob deformation, speed, and elongation. Figure 5 is an example of the normal forces for the current generation of delivery equipment compared to an optimized geometry. The old geometry is on the right and the sharp increase in force can be seen, for the optimized geometry the increase in force is much smoother.

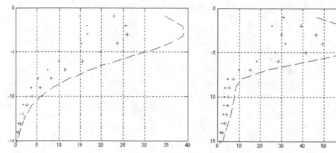

Figure 5: Normal force load

Computational fluid dynamics (CFD) simulations were conducted to further validate the optimized scoop trajectory generated by employing the Bezier curve generation and normal force analysis approach. A good agreement was seen between normal force analysis and the CFD predictions.

It was further hypothesized that by forcing the gob through a reduced cross section in the scoop it will extrude the gob to a very repeatable length, as long as the weight and gob viscosity remain constant. The other benefit is that the gob can not wander down the scoop. From tests it has been seen that, in some cases, the gob actually slaloms down the scoop. This leads to a landing variation in the trough, and eventually ends up as container defects due to an oddly shaped gob.

A new scoop has been designed to evaluate the use of a reduced exit cross section and an improved trajectory. The new scoop cross section design can greatly reduce the variations in the gob due to the scoop or feeder. With proper cross sectional areas it will be possible to improve the gob shape, maintain gob speed, and reduce the variation on the trough. Another effect of the new scoop will be to fix the 'dog-bone' shape of the gob. This can be seen from the CFD results, in Figure 6, as well as validated by high speed video.

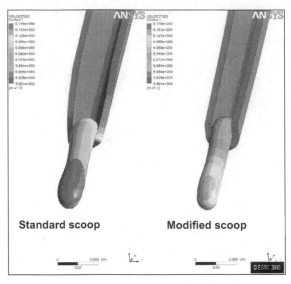

Figure 6: Gob shape comparison

From test results, the new scoop combining the trajectories and the new cross section designs meets all the requirements that were set out to achieve: An increase in velocity, a better and uniform gob shape, and a negligible increase in length.

AIR JET TROUGHS

The next impact point is the entrance of the trough. The majority of troughs are now fabricated from steel and coated with either an epoxy based dry coating or plasma sprayed coating. The effectiveness of these coatings is short-lived at the gob impact area of the trough. Epoxy coatings tend to wear through in a few weeks and plasma coatings gradually lose their effectiveness.

The integration of a small nozzle in the gob landing zone of the trough, as shown in Figure 7, provides an air layer that protects the gob from the impact with the trough. The input air has the effect of cooling this landing region and more importantly providing an air cushion which reduces the impact, thus decreases damage to the gob and minimizes the variation in the final load to the blank mold. The air cushion also helps extend the coating life at the impact region due to the reduction in friction at the landing zone.

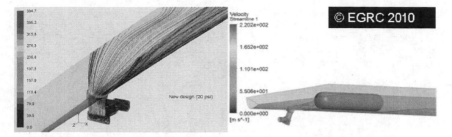

Figure 7: Air cushion in gob landing zone of trough

Figure 8 shows the measured variation in gob velocity in the middle of trough. The steel is coated with a standard epoxy dry delivery coating, and measured at the time of the trough had been in operation for 10 weeks. As can be seen from the plot, the air was turned off to the nozzle for a period, resulting instantly in a significant increase in gob speed variation, as well as slower gob speed. As soon as the air input is turned back on, an instant recovery and variation of less than 0.2m / second were observed in the test results.

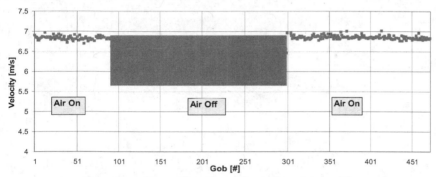

Figure 8: Practical confirmation of effect of air nozzle in trough

CONCLUSION

Transporting the gob from the shears to the blank molds can be a difficult process leading to a variety of issues affecting the final bottle quality. Using a newly developed system to monitor the gob as it travels through the delivery system has helped us gain insight into what causes these problems and where they occur. In an attempt to mitigate the variations caused by the various components of the delivery system each one was optimized or reinvented. Utilizing a gob stabilizer prior to the scoop we are able to guarantee that the gob enters the scoop vertically this fixes any length and velocity issues the gob may have if it impacts the scoop in the wrong place. A scoop with a tapered cross section is used to force a predetermined shape upon the gob. Knowing that the gob will keep its general shape throughout the delivery system this becomes important because we can ensure that the gob is the proper shape to load into the blank mold. The transitions from the scoop to the trough and the trough to the deflector cause a large impact on the gob leading to elongation and a loss in velocity, by using an air cushion at the entry regions this effect can be mitigated and the coating life extended. By addressing these problems at their root cause, it becomes possible to deliver a better gob to the blank molds and thus create a bottle with less delivery related defects.

Controls and Raw Materials

MODEL BASED PROCESS CONTROL FOR GLASS FURNACE OPERATION

Piet van Santen, Leo Huisman, Sander van Deelen
TNO Glass Group
Eindhoven
The Netherlands

ABSTRACT

Since the eighties of the 20th century supervisory control systems have been used in the glass industry. Model Predictive Control (MPC) technology, which was developed in the chemical industry, was introduced in the glass industry for control of forehearths and crown temperatures in glass melting tanks. The mathematical (black box) models used in these controllers were determined from process data. To this end tests were performed on the process so that the time dependent behavior of the process could be observed and the model parameters could be estimated.

In the last ten years application of Computational Fluid Dynamics (CFD) models has become possible through model reduction. Via Proper Orthogonal Decomposition the relevant flow and temperature characteristics in the glass melt can be found for the normal operating range of the process. Depending on the operating range linear or nonlinear reduced models can be chosen to make fast predictions of the process behavior.

This paper presents some recent results of industrial applications of MPC based on reduced CFD models, called the RMPC approach. Results will be shown for a container glass forehearth. The forehearth controller was designed for two glass colors and two models were needed to cover the working range. In both applications the variation of the relevant temperatures was reduced by 50% or more.

INTRODUCTION

Process control for glass melting furnaces is important to operate the furnace at lowest energy consumption, lowest possible disturbances affecting glass quality and to improve process stability even for variable input (e.g. cullet and pull variations).

The main purpose of supervisory control systems for glass furnaces is to provide optimal set-points to the primary control parameters (fuel input, batch composition, air/gas ratio) to produce glass with minimal energy and with highest quality even at changed conditions. These systems are installed on top of existing primary control systems.

The supervisory control solutions are based on industry standard Model-based Predictive Control technology (MPC). This widely accepted technology uses a dynamic process simulation model of the furnace (calculating temperatures and sometimes also glass melt flows, residence times and even glass melt quality parameters) to predict the future process behavior and to adapt its control actions continuously to compensate for observed disturbances.

The quality of the controller performance is largely determined by the accuracy of the simulation model. If the prediction is less accurate, then the control system will run with lower performance. It has to cope with disturbances from the outside world, but also with the model errors.

In this paper we present a supervisory control system based on a detailed numerical process model using GTM-X. GTM-X is the name of the numerical 3D Furnace Simulation software from TNO. GTM-X is one of the applications of the OSS (Operation Support System) technology of TNO.

Until 2003 GTM-X was mostly used off-line by corporate R&D and engineering for trouble-shooting, process optimization and evaluating (re-)designs. The simulation speed was the limiting factor for on-line applications. TNO has developed a model reduction technology (see [4]) which makes the model 1.000 to 10.000 times faster with sufficient model accuracy for Model-based Predictive Control (MPC).

Main attention in this paper will be the differences in the approach between the conventional MPC systems based on step testing and new MPC systems based on detailed numerical models (RMPC). The modeling approach and control strategy will be demonstrated on an industrial container glass feeder and obtained results will be given.

CONTROL STRATEGY

Most glass melting furnaces and forehearths in the container glass industry are still controlled using conventional PID controllers supervised by operators. In forehearths typically one or more temperatures are selected per section, each of which is controlled by a single loop (PID) controller (Figure 1a). Sometimes the average of a few sensors is taken as the controlled value. If the glass temperatures at the end of the forehearth deviate too much from their desired value then the operator adjusts the setpoints of the section temperatures that are controlled by the PID controllers. In practice the setpoints are rarely changed between job changes and if they are changed then the operator selects one or more setpoints which he then changes step wise.

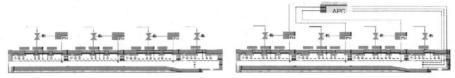

a. Conventional PID control   b. Supervisory controller (Advanced process control)
Figure 1: Conventional PID control and Advanced Process Control (APC) of a typical forehearth

In the last few decades multivariable supervisory controllers have been applied on industrial forehearths ([1],[2],[3]). Such a controller plays the role of the supervising operator as it manipulates the setpoints of the PID controllers in order to maintain a desirable temperature profile in the spout section (Figure 1b). The temperature profile in (or just before) the spout section is measured by a 9 point grid measurement (three triple elements through the superstructure).

Model Predictive Control (MPC)

The supervisory controller used for control of the entire forehearth is multivariable, should be able to deal with dead times and needs to respect limits on all relevant variables. A Model Predictive Controller (MPC) is well suited for control of such systems. An MPC uses the available online process measurements to predict the future process behavior and find the optimal future manipulations of the setpoints for the PID controllers (manipulated variables, MV) to obtain the desired future behavior of the controlled variables (CV). In [5] a survey is given of the currently applied MPC technology on the market. This is illustrated in figure 2. If other influences on the process are known the MPC can take these into account in the predictions. Examples of such known disturbance variables (DV) are the pull rate and the glass temperature at the inlet of the forehearth.

In this application the INCA controller of IPCOS is used. This MPC controller is part of a software architecture that runs on a standard PC and communicates with the DCS system via an OPC connection. OPC is a standard communication protocol for process control.

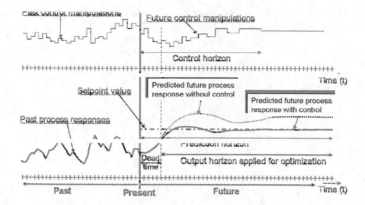

Figure 2: Model predictive control principle: Given the past and present information, find the future control moves (over the control horizon) such that the predicted future process response is as close to the desired response as possible and constraints are not violated.

The forehearth discussed in this paper has 4 control sections and is equipped with 27 thermocouples and 12 local controllers. On the left (L) and right (R) side groups of burners are used to control the left and right top temperatures in the section. In the center cooling air is used to control the center temperature in the section (Figure 3).

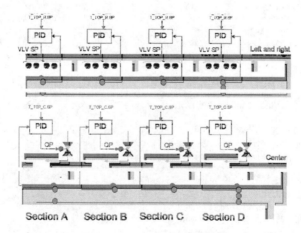

Figure 3: Control scheme of the feeder. The feeder has 12 control valves and 21 thermocouples that are relevant

Six of the 27 temperatures were not used for control, so the number of relevant temperatures was 21. In addition to these the inlet temperature is measured just before the first section in the center and also the pull rate is manually entered whenever there is a job change. These latter two variables are used as DV in the controller. This means that in principle (12+2)*21 = 294 transfer functions have to be determined to describe the relations between all inputs and outputs. Fortunately, there are zero transfers that do not have to be estimated since the PID controllers do not influence the upstream sections. However, the number of model transfers that needs to be determined is still quite large.

SIMULATION MODEL

Conventional MPC

Before 2003 MPC systems were applied in the glass industry based on conventional data-driven identification methods. To obtain data for the modeling procedure, step tests or PRBNS (Pseudo Random Binary Noise Signal) tests are performed on the feeder to be controlled.

The model derived from these tests is only valid for the operating window covered during the tests and can only provide good control performance within this operating window. When the process situation changes (e.g. different pull, different glass composition, or different fuel) the data-driven derived model loses its validity and does no longer provide adequate on-line performance. Therefore, it is desired that a large enough operating range is covered by the tests.

For the feeder discussed in this paper typically 3 – 4 weeks of testing are needed to obtain a sufficiently large amount of data that enables a complete model fit. Note that 12 inputs have to be manipulated in order to get data for the mathematical model. As will be argued below the testing procedure has to be repeated if another glass color is being produced.

Rigorous model based MPC (rMPC)

As stated in the previous section for control it is desirable to have a model that best represents the process behavior in the expected range of operation. Besides the fact that only a small part of the working range will be available at the time of testing, also process disturbances are present that make it more difficult to extract the process behavior from the data. For example, when step tests are performed when at the same time there are variations in cullet contamination these variations will disturb the process.

To obtain good data for model identification, rather large steps are necessary on the process inputs. In case large changes are not allowed, changes have to be applied for a longer period in order to filter out the effects of disturbances. The rigorous model contains no disturbances and there are no limits on changes on inputs.

The controller can react via feedback to unknown disturbances as soon as their effect is observed in the thermocouple values. If the disturbances are known (measured) the controller can pro-actively anticipate their effect through feedforward control.

Some disturbances are measured and can be taken into account in the controller. An important example is the pull rate. In tests on the process, however, the pull rate can hardly be taken into account for at least two reasons:
- One generally is not allowed to change the pull just for testing purposes.
- If the pull is changed for production purposes, often other variables are changed at the same time, making it hard or impossible to extract its effect on the process from the data.

To avoid time-consuming tests on the real process, with the described drawbacks, TNO developed the RMPC approach ([1]). In this approach fast reduced simulation models are derived from a detailed first principles model GTM-X. This Computational Fluid Dynamics (CFD) code can handle body fitted grids which is necessary for accurate simulation of the complicated geometry (see Figure 4).

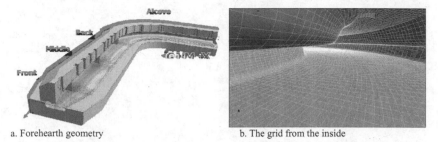

| a. Forehearth geometry | b. The grid from the inside |

Figure 4: The geometry and the grid that were used in the GTM-X model of the forehearth.

The spatial domain was divided into $1.5 \cdot 10^6$ grid cells in 6 sub domains: the glass melt, the walls and the combustion spaces in the 4 sections. In the curved alcove section the temperature and flow pattern become asymmetric. This can be visualized by a particle trace in the forehearth. Figure 5 shows the released particles (50000) after 20 minutes. The particles on the left side have a shorter path to travel and

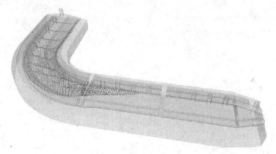

Figure 5: The geometry and the grid that were used in the GTM-X model of the forehearth.

From the GTM-X model a reduced model was derived. To this end step tests were performed using the GTM-X model. The forehearth is used for two glass colors: green and flint. For both glass types ~ 4 weeks of process time were simulated with the GTM-X model. From the collected data a reduced model was derived.

The reduced model described the static and dynamic relationship between the gas and cooling air flows in all zones and the glass temperatures. The step responses of the 9 point grid in the front section are shown in figure 6. To be able to use this model in the MPC the existing local controllers needed to be incorporated in the model as well. First, these controllers had to be retuned because some control loops had a slow response. The setpoints of the top temperatures in all sections were used as manipulated variables (MV) in the RMPC controller.

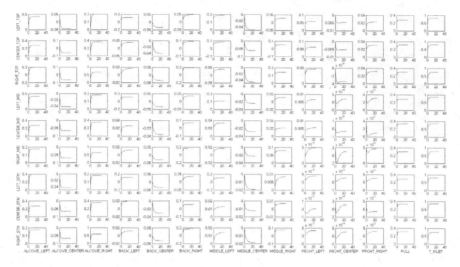

Figure 6: Step responses of the 9 front temperatures in reduced model

Effect of glass color

The transmission properties of the two produced glasses differ significantly. This greatly affects the (dynamic) behavior of the glass temperatures. Figure 7 illustrates the difference in response between the two colors.

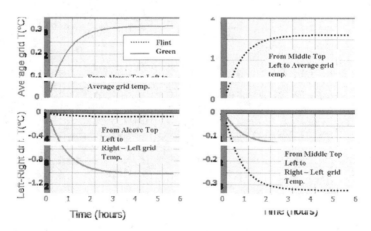

Figure 7: Some step responses for flint glass and green glass.

A setpoint step in the alcove section has a significant effect on the front section temperatures when green glass is produced but the effect for flint glass is small. This can be explained by the fact that the increase of energy put into the glass in the alcove section is compensated by a decrease of energy in the back and middle section in order to keep the top temperatures at their setpoint (see figure 3). Since the radiative heat transfer is larger for flint glass, the control actions in the subsequent sections will have a larger effect on the bottom glass. This causes the effect of the setpoint change in the alcove section to be suppressed more for flint than for green. On the other hand a setpoint change in the middle section has a smaller effect for green glass for the same reason. The middle section is too close to the front section for the controllers to affect the bottom glass.

Since the responses for green and flint glass are significantly different one linear model will not be sufficiently accurate for both glass colors. Therefore, two controller models were derived, one for each color. In case of the traditional black box approach, two test periods would have been necessary.

CONTROLLER RESULTS

The RMPC controller uses the setpoints of the top temperatures in the forehearth sections to control the 9 point grid in the front section. The top controllers in the front section are not used in the RMPC. They are used to keep the top temperatures in the front section at their setpoints.

Instead of controlling the individual temperatures, the RMPC controls the average over the 9 thermocouples and the average temperature difference between right and left (Figure 8).

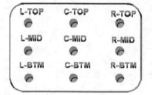
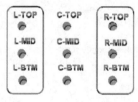

a. Average of 9 point grid measurement   b. Average right temperatures minus left temperatures

Figure 8: Controlled variables used in the controller

Note that the top temperatures in the front section are fixed by the local controllers (Figure 3). So, effectively the other six are influenced via the two CVs.

Table 1: Standard deviations for Green glass	σ (Without RMPC)	σ (With RMPC)
Average of 9 temperatures	1.89	0.32
Delta T (Right – Left)	1.93	0.69
Delta T (Top – Bottom)	4.21	0.79

Table 2: Standard deviations for Flint glass	σ (Without RMPC)	σ (With RMPC)
Average of 9 temperatures	1.04	0.11
Delta T (Right – Left)	1.57	0.21
Delta T (Top – Bottom)	0.13	0.19

Figure 9 shows some collected data of the two CVs without control and with RMPC control. Tables 1 and 2 show the standard deviations of the CVs for green and flint glass (for longer periods than shown in figure 9) before and after implementation of the RMPC controller. The variation of the controlled variables is significantly reduced for both glass colors.

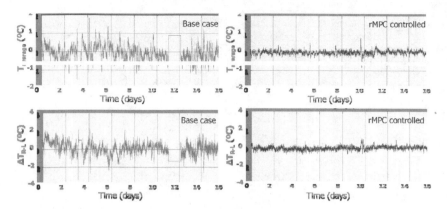

Figure 9: Controlled variables before and after implementation of RMPC

In tables 1 and 2 also the results are shown for the temperature difference between top and bottom. Variations of this temperature difference are only reduced in the green glass case. In case of flint glass the variation without control is already small and no improvement is observed.

The shown results can only be obtained if the controller actually has enough freedom to keep changing the top temperature setpoints. If too many temperature limits become active in the alcove, back and middle section then the controller can no longer achieve the setpoints for the average temperature and the right-left temperature difference nor can these CVs be stabilized.

Besides more stable temperatures in the front section the RMPC controller helps in reducing the reject rate and the number of operator interventions. In this case the operators of the forehearth are also responsible for the forming process. With a properly working RMPC the operator has more time available to perform this latter task. The glass manufacturer has recognized the advantages of the RMPC controller and similar controllers will be implemented on the other forehearths and on the melter.

CONCLUSIONS AND FUTURE DEVELOPMENTS

In this paper we have presented the results of a recent RMPC project for a container glass forehearth. The RMPC approach was described and the essential differences with conventional MPC were presented. Where conventional MPC relies on (step) tests on the forehearth at a convenient production period, RMPC is based on reduced CFD models which can be constructed for an arbitrary large operating range without disturbing the process. It was shown that different controller models were needed for the two produced glass colors. In that case the conventional approach would require tests during a green glass production period and during a flint glass production period.

The controller results were presented. It was shown that the variations on the average of the 9 point grid temperatures and the difference between right and left were reduced for both glass colors. It can be concluded that the RMPC can be successfully used to stabilize the forehearth glass temperatures without the need to perform tests on the process.

TNO keeps investing time in further improvements of the RMPC technology and developing new technology. Currently, for example, fast nonlinear 3D reduced simulation models have been developed for velocity fields in a container glass melter and effort is made to develop soft sensors based on these models to estimate velocity field characteristics such as back flow in the throat.

Furthermore, a research project is proposed to study and improve the controllability of feeders and to improve gob temperature stability. To this end the GTM-X model will be extended with the TNO spout model so that relationships between the 9 point grid temperatures and the gob temperatures can be determined and taken into account in the RMPC controller.

REFERENCES

[1]   Ton Backx, Leo Huisman, Olaf op den Camp, Oscar Verheijen, Rigorous model based predictive control of a glass melter and feeder, Glass Technology, Volume 49 (3), 139-144, June 2008

[2]   J. Chmelar, R. Bodi, E. Muysenberg. Advancing control of glass melters and forehearths, Glass, July 2000

[3]   L. Huisman, S. Weiland. Identification and model predictive control of an industrial glass feeder, In preprints IFAC symposium on system identification, Rotterdam, August 2003

[4]   Leo Huisman. Control of glass melting processes based on reduced CFD models, PhD thesis, Eindhoven, March 2005

[5]   S.J. Qin, and T.A. Badgwell, A survey of industrial model predictive control technology, Control Engineering Practice, Nr. 11, 2003, 733–764

TAKING FULL BENEFIT OF OXYGEN SENSORS AND AUTOMATIC CONTROL

Peter Hemmann
STG-Software & Technologie Glas GmbH Cottbus (Germany)

Since more than 20 years STG has been active in production, installation and services for oxygen sensor applications to the glass industry. These sensors became a common standard with a service lifetime of about 3 years, best results up to 8 years, applicable up to 1500 °C, applicable even under highly reducing atmosphere, providing a signal good for reliable measurement and good for automatic control. About 10…20% of the applications are really using them for automatic control.

Figure 1: Oxygen sensor in regenerator crown installation

We install them typically in the regenerator crown in vertical position – which gives better service life and never gives any problems with measuring accuracy due to this position. All this would not be a reason to come over the ocean for giving this paper. New developments are really going on in the signal processing of these sensors, in the answer to the question – which value can such sensor give you?

Oxygen sensors give you much more information than oxygen percentages only. Combined with the information about fuel composition – may be gas or oil or both in a mix – the system can make the full combustion calculation.

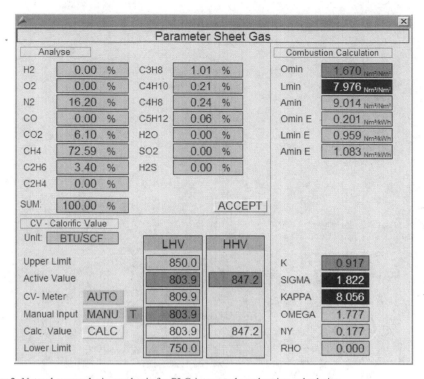

Figure 2: Natural gas analysis as a basis for PLC-integrated combustion calculation

Another point to be observed is the gas composition. With our experiences the gas composition is not constant. If Calorific value varies – so stoichiometric air demand and dimensionless gas parameters K, SIGMA, KAPPA etc. will be vary as well.

If you have available an actual process value of natural gas calorific value, so the software block shown in Figure 2 will adjust the key figures accordingly.

Based on the actual gas composition, the signal processing provides the amount and the composition of flue gas. Providing CO percentage requires some more comment.

We get from sensor voltage not only Oxygen percentage, but based on Boudouard chemical balance the CO percentage corresponding to $O_2\%$ - in chemical balance only. This is the basis to get the Lambda excess air value

$$\lambda = 1 + \frac{0,2094*(\kappa+\omega+Ge)-0,7906*\psi}{\sigma} * \frac{[O_2\%]}{20,94\%-[O_2\%]} - \frac{\psi}{\sigma}$$

notice: 20,94% is oxygen percentage in oxidant (air) $O_2\%$ in Vol.%
79,06% is nitrogen percentage in oxidant (air) N_2 in Vol.%

Using abbreviation as

(D_46) $$\psi = \frac{(1+Ge)*RaCO}{2*(1+RaCO)}$$

Oxygen sensor signal processing will not give you any CO temporarily exceeding the chemical balance figures due to turbulent combustion – but it gives you reliable information how much excess air or how much air is missing in the giving combustion.

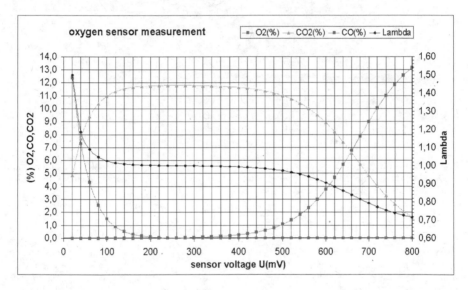

Figure 3: Oxygen, CO and Lambda as a function of sensor voltage for given temperature and for given natural gas composition

Roughly to say: For sensor voltage U < 200mV sensor primarily indicates you Oxygen percentage, CO is quite low – in chemical balance.

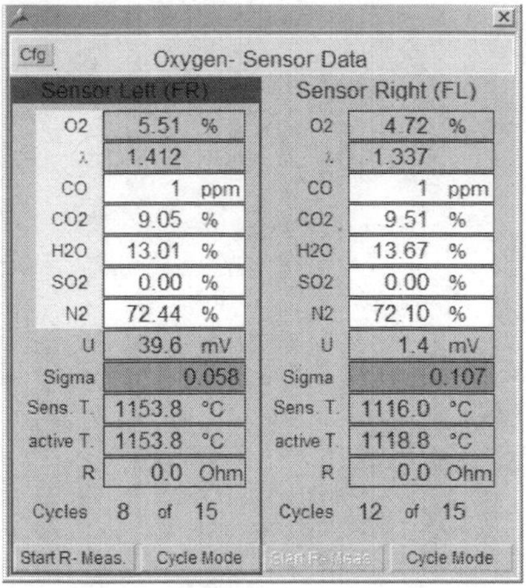

Cfg	Oxygen- Sensor Data		
Sensor Left (FR)		Sensor Right (FL)	
O2	5.51 %	O2	4.72 %
λ	1.412	λ	1.337
CO	1 ppm	CO	1 ppm
CO2	9.05 %	CO2	9.51 %
H2O	13.01 %	H2O	13.67 %
SO2	0.00 %	SO2	0.00 %
N2	72.44 %	N2	72.10 %
U	39.6 mV	U	1.4 mV
Sigma	0.058	Sigma	0.107
Sens. T.	1153.8 °C	Sens. T.	1116.0 °C
active T.	1153.8 °C	active T.	1118.8 °C
R	0.0 Ohm	R	0.0 Ohm
Cycles	8 of 15	Cycles	12 of 15
Start R- Meas.	Cycle Mode	Start R- Meas.	Cycle Mode

Figure 4: Indication of flue gas composition, Lambda value and Sigma turbulence indicator

But going further into reducing situation, we see sensor voltage increasing more and more up to 800 mV, where Oxygen is more or less no more measurable, but in this area U >200mV sensor voltage primarily becomes an indicator for increasing percentage of CO.

Lambda excess air figure is available over the full range of sensor voltage, indicating oxidizing and reducing conditions as well. Please notice, that Lambda is a real process value, representing the status of total air which was available for the combustion. To get it we need just only the sensor voltage, temperature and fuel composition.

Figure 4: Indication of flue gas composition, Lambda value and Sigma turbulence indicator

Sigma here represents the standard deviation of sensor voltage – divided by the average of the voltage – to be used as an indicator for the turbulence of combustion, which becomes interesting based on the general knowledge, that the less turbulent the combustion is, the better will be efficiency and low NO_x results.

Now let us come to the most interesting point: sensor application for automatic control.

Lambda excess air value is much better qualified for automatic control function, than oxygen percentage is. This is due to two reasons:

a) Lambda value covers whole scope of possible sensor voltage, oxidizing and reducing combustion by same algorithms and same way of control

b) Lambda is in a linear relationship with combustion air – therefore any required difference in Lambda may be directly converted into required modification of air flow, without waiting for the response of the combustion system

Lambda process value PV_Lambda gives us the amount of uncontrolled air or ingressed air PV_XF as:

$$PV\_XF = PV\_Lambda * LMIN * PV\_fuel - PV\_comb.air$$

(LMIN = stoichiometric demand of air)
Uncontrolled air XF is the major concern of automatic control – we call it Lambda Control. Lambda

Sources of air ingress

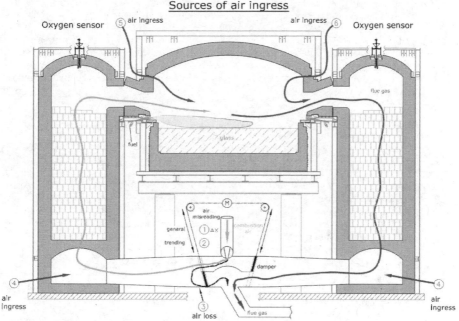

Control means:

- Monitor ingressed air
- Minimize ingressed air
- Indicate ingressed air's source
- Compensate ingressed air – whatever is not avoidable and whatever is acceptable to compensation

Figure 5: Sources of uncontrolled air XF

There are different sources of uncontrolled air – requiring different control strategies to compensate or to deal with:

(1) General misreading of combustion air measurement "appears like" uncontrolled air – is not critical, but better to find out by comparison of two or more operation points at different fuel flow

(2) Drifting misreading of combustion air flow – comes with an increasing amount of "hidden air" and this is fully correct to be seen and compensated as air ingress

(3) Loss of combustion air – air flow short-circuit caused by incompletely closed reversal damper – gives a negative air ingress and can be identified by typical trending behavior. Air loss is typically increasing over the duration of firing period – so oxygen reading gives a falling trend

(4) Air ingress into regenerator bottom where the lowest pressures are on the combustion air side – is a typical air ingress and can easily be compensated by modification of combustion air flow.

(5) air ingress into furnace chamber not taking part in heat up in the regenerator or even

(6) air ingress into furnace chamber even not taking part in combustion

Any uncontrolled air of types (1), (2), (3) or (4) can be fully compensated by modification of combustion air flow where the new setpoint is given by:

$$SP\_Air = SP\_Lambda * LMIN * MAX(PV\_fuel, SP\_fuel) - XFA$$

Where XFA is a figure for the ingressed air, but using not just the actual process value, better taking in account the typical trending patterns of the last 5 reversal periods.

Compensating any ingressed air of types (5) and (6) – entering into furnace chamber – will risk to disturb thermal balance of regenerators and can be done in close limits only.

Therefore it becomes an important point, how to make a difference between air ingress into bottom of regenerator or hidden air from combustion air misreading on the one side – and air ingress into furnace chamber on the other hand.

Answer is: we use the temperature footprint of ingressed air in the regenerators, to identify whether it can be compensated by modification of ingressed air or not.

There is a software block to check on-line energy balance of the regenerators:

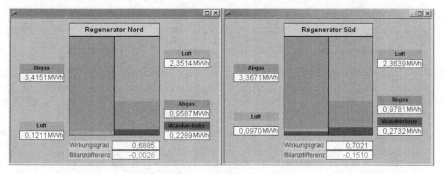

Figure 6: On-line energy balance of regenerators

The different types of ingressed air have different impact on the energy balance of regenerators:

Type (1) and (2) affect both firing sides in the same way and can be compensated by modification of combustion air without any affect to the energy balance of the regenerators.
Type (3) and (4) of uncontrolled air – air loss or air ingress into regenerator bottom – will have an effect different between the firing sides – and compensating them by modification of combustion air will just improve thermal balance of regenerators – providing finally same temperatures at both regenerator crowns.
Worst cases are type (5) and (6) of air ingress – stealing energy from the furnace and disturbing thermal balance of regenerators. Any effort to compensate direct air ingress into furnace will result in a further disturbance of thermal balance of left and right regenerator and is very likely to increase regenerator temperatures, which is again not a welcome result.

The only way out is: seal the furnace and increase furnace pressure.

So conclusion is: Identify ingressed air, make it a "normal" process value.
Compensating it by modification of combustion air flow works perfectly as long as the reason of air ingress is anywhere in regenerator bottom or upstreams of regenerator. In these cases compensation even will improve thermal balance of regenerators.

Compensating of ingressed air entering directly into furnace chamber disturbs the thermal balance of regenerators and cannot be done by modification of combustion air – or at least only in very close limits – increase of furnace pressure is the more reasonable way.

This is what we call "the temperature footprint of ingressed air"

Compensation of ingressed air by modified combustion air works perfect in most of the situations. But we have to accept, that there are situations, where it cannot work – just when it would be disturbing thermal balances and increase regenerator temperatures.

Practically we use an automatic control of thermal balance of regenerators, playing actively – but in limits – with a shifting time between left hand side and right hand side firing period to transport heat from the hotter to the colder regenerator in order to get same temperatures or better – in order to get same amount of preheated air energy flow into furnace from both regenerator sides.

All these are slow control processes. Balancing the regenerators may take days, even a week when starting from an unbalanced situation. Later on shifting time comes down to nearly zero – which indicates that compensation of ingressed air is correct without disturbing the thermal balance of regenerators.

Or: Over days and weeks it will be found that balance controller requires a more or less stable difference between left side and right side period – indicating that there is a continuous reason for thermal unbalance of regenerators, resulting eventually from ingressed air into furnace compensation. In such situation, an increase of furnace pressure should help to reduce ingressed air and to improve thermal balance of regenerators.

Showing how a well functioning Lambda Control should be working:

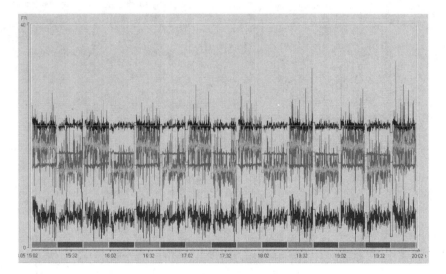

Figure 7:   Lambda Control at an endport furnace
(red = ingressed air, yellow XFA to compensate, dark blue is oxygen O2%
black and clear blue Lambda PV and SP, green is ratio slightly different left and right)

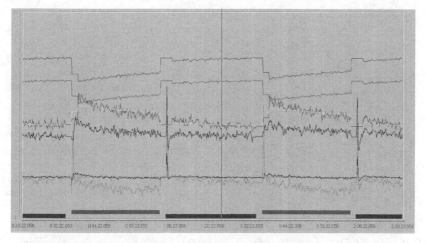

Figure 8:   Lambda Control at a crossfired furnace having 2 air flow control groups
(red = ingressed air, yellow XFA to compensate, dark & light blue is oxygen O2%
black and clear blue Lambda PV and SP, green1&green2 is ratio slightly different left and right)

Figure 8 is showing Lambda Control for a cross-fired furnace, having 5 ports and only 2 oxygen sensors and 2 groups of air flow control.

Lambda Control has to take care for the neighboring ports also, with no oxygen sensors, calculating limits to consider the eventually hidden risk of too low Lambda value on one of the neighboring ports.

The diagram shows decreasing trend patterns of ingressed air and oxygen – resulting in increasing trend patterns of ratio, especially for left fire (see fire side indication on bottom line) which indicates a possible air leakage of reversal dampers for left fire.

CONCLUSION:
Lambda Control has become a tool worth to be used for efficient furnace operation. We have learned to "read" indications of misreading and air losses, we have learned carefully to see not only oxygen sensor results but also the temperature footprint of ingressed air. And we know that there are limits also for the compensation of ingressed air by modified combustion air – limits set by the requirements of thermal balance of regenerators and furnace pressure.

FLUE GAS TREATMENT IN THE GLASS INDUSTRY: DRY PROCESS AND CALCIUM-BASED SORBENTS

Amandine Gambin and Xavier Pettiau
Lhoist Group, Nivelles, Belgium

1.   SUMMARY

Among the different flue gas treatment processes, the injection of dry sorbent fits particularly well with the needs of the glass industry. The benefits using dry sorbent injection include technical simplicity, low overall cost for the investment, and low maintenance costs for flue gas treatment. *Lhoist* offers Sorbacal® SP, a specialized porous hydrated lime product specifically designed to reduce sorbent consumption while reducing waste residue production. This product further enhances the benefits of utilizing dry sorbent injection compared to other methods. Additionally, in order to address increasing market demand for flue gas desulfurization, *Lhoist* has also developed Sorbacal® SPS, a porous hydrated lime that exhibits enhanced desulfurization performance.

The flue gas removal performance of these sorbents has been studied in a flue gas pilot plant designed to simulate operating conditions found in the glass manufacturing industry. These results compare favorably to those obtained during an actual industrial trial performed at a glass factory.

2.   INTRODUCTION

The glass making process releases quite high levels of acidic gases, such as sulfur dioxide (SO_2), considered the major pollutant, hydrochloric (HCl) and hydrofluoric (HF) acids. In order to remove these pollutants in the context of glass industry legislation, several flue gas treatment methods are available and grouped in three categories: wet, semi-wet and dry processes [1]. Among them, the dry process, which consists of injecting a powdered sorbent typically calcium hydroxide into the gas, combines low levels of investment and maintenance costs, simplicity, robustness and absence of liquid effluent [2,3]. For these reasons, it fits particularly well with the needs of the glass industry.

Nevertheless, the main draw-back of this process concerns the gas/solid reaction leading to an over-consumption of reagents and consequently an over-production of solid by-products [4]. It is for this reason Lhoist developed new reagents [5] such as Sorbacal® SP, a highly porous hydrated lime which offers removal performance proven to be greatly above standard calcium hydroxide [3, 6]. Sorbacal®SP differs from "standard hydrated lime" products in that it has very high BET specific surface area (>40 m^2/g) as well as pore volumes that are 2-3 times those of typical hydrated lime products. Additionally, to address an increasing demand for high desulfurization levels, Lhoist recently developed a new product called Sorbacal® SPS which is an activated porous calcium hydroxide with enhanced desulfurization performance.

This paper presents a parametric study of the removal efficiency of Sorbacal® SP and Sorbacal® SPS on a pilot scale under operating conditions consistent with those found in the glass industry. This approach is corroborated by the results of an industrial scale trial carried-out in the glass sector.

3.   DESULFURIZATION PERFORMANCE ON A PILOT SCALE

The aim of this pilot scale study is to measure and compare the acid gases' (SO_2 and HCl) removal performances of calcium-based sorbents under controlled operating conditions. In an effort to to be as realistic as possible, these trials have been performed with a gas mixture composition similar to the one typically experienced in the glass industry.

3.1.   Description (see Picture 1)

The pilot scale installation is composed of a tubular reactor that is 4 centimeters in diameter and ± 3 meters tall. The calcium-based sorbent is injected co-currently to the acid gases from the top to the bottom, utilizing a powder-feeder swept by nitrogen. The facility has four lines equipped with flow-rate regulators delivering controlled amounts of HCl, SO_2, CO_2 and N_2 as well as a water vapor generator. The total gas flow-rate ranges between 1000 and 2000 Nl/h. The acid gases' concentrations are continuously measured at the reactor outlet by an infrared analyser. During each test, the gases are pre-heated before entering the reactor and the tube itself is maintained at a constant temperature. The sorbents' residence times are in the range of seconds.

Picture 1. Pilot scale installation

3.2.   Operating conditions

The glass industry flue gas compositions vary from one site to another. However, SO_2, which is considered the major acid pollutant, is present at levels ranging between 500 and 2000 mg/Nm³ at 8% O_2. The HCl and HF contents are significantly lower (about 5 to 100 mg/Nm³) [7].

Consequently, the sorbents' performances were studied for the removal of SO_2 alone as a function of temperature (ranging between 150 and 500°C), moisture (0, 5, 10 and 15%) and CO_2 (0 and 9%). The inlet SO_2 concentration has been set to 1500 mg/Nm³.

3.3.   INFLUENCE OF TEMPERATURE AND MOISTURE ON SORBACAL®SP PERFORMANCE

Figure 1 presents the SO_2 removal rates reached by Sorbacal®SP at a feed rate of 7 g/h with 9% of CO_2.

We observe three operating zones:

➤ Below 200°C, the removal performances decrease in the case of dry gases and increase drastically as soon as moisture is present. This confirms the significant influence of relative moisture on SO_2 abatement.

➤ Between 200 and 350°C, performance increases with temperature. The moisture influence becomes negligible.

➤ Above 350°C, the Sorbacal®SP performance decreases with increasing temperature. At 500°C, moisture has a positive effect on SO_2 removal because it delays the decomposition of calcium hydroxide into calcium oxide.

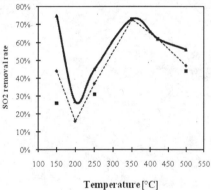

Figure 1. Influence of temperature and moisture on the Sorbacal®SP desulfurization performance on a pilot scale.

3.4.   Comparison with Sorbacal®SPS

A comparison between the removal efficiencies reached by Sorbacal®SP and Sorbacal®SPS under the same operating conditions (9% of CO_2, 10% moisture, feed rate 5 g/h) demonstrates that using Sorbacal®SPS enables performance improvements by at least 15 to 20% at temperatures between 150 and 350°C (see **Figure 2**.)

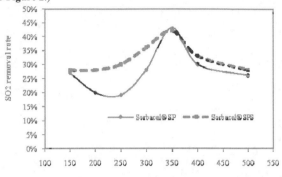

Figure 2. Sorbacal®SP and Sorbacal®SPS desulfurization performance on a pilot scale

3.5.   Comparison with a standard hydrated lime

The comparison between the performance of a standard hydrated lime (such as Sorbacal®H) and Sorbacal®SP, expressed as a calcium sulfite transformation rate determined for a given SO_2 removal of 50%, confirms with 25% of moisture and 9% of CO_2, better efficiency is noted when utilizing Sorbacal®SP at temperatures below 450°C (see **Figure 3**). Evolution of the removal rates of both sorbents follows the same trend up to 350°C. Above this temperature, the Sorbacal®SP performance decreases while standard hydrate performance continues to increase. At 500°C the Sorbacal®H performance is better than that of Sorbacal®SP.

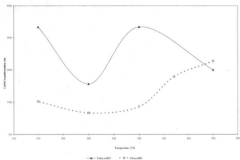

Figure 3. Sorbacal®SP and standard hydrated lime desulfurization performance on a pilot scale.

3.6.   Influence of the CO_2 gas content

Figure 4 presents the SO_2 removal achieved by Sorbacal®SP with 9% CO_2 as compared to 0% CO_2 at 25% moisture and a feed rate of 7 g/h. Above 150°C, CO_2 degrades the performance of this sorbent. This effect is accentuated with increasing temperature, especially above 350°C, where the removal rates tend to stabilize without CO_2 and then start to decrease with CO_2 present in the gases. This result suggests that the competition between the reaction of Sorbacal®SP with SO_2 and its carbonation is exacerbated beyond 350°C.

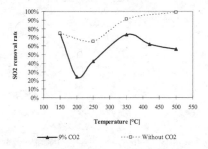

Figure 4. Influence of CO_2 on the desulfurization performance of Sorbacal®SP on a pilot scale.

3.7.   Conclusion

This parametric study demonstrates that the desulfurization performance of hydrated limes is strongly influenced by operating conditions (temperature, moisture, CO_2...). In fact, each case needs to be studied before choosing a specific sorbent. At temperatures above 500°C, Sorbacal®H is recommended; between 350 and 500°C, Sorbacal®SP is the best choice and below 350°C, Sorbacal®SP or SPS can be used depending on the required removal level.

It is important to note that this pilot scale study cannot reflect the complexity of a real industrial situation, where the combined removal of SO_2, HCl and HF takes place. Due to these complexities, this study requires confirmation by an industrial case study in the glass industry.

4.   INDUSTRIAL SCALE TRIAL

4.1.   Trial objectives

This trial was carried-out in a European glass making factory. This plant was equipped with a dry process fed by sodium carbonate and was not able to meets its required HF emission limit.

The goal of this trial was then to determine the minimum Sorbacal®SP feed rate needed to achieve the required HF level (below 5 mg/Nm³ at 8% O_2). Additionally, the desulfurization performances of this sorbent were also studied considering different SO_2 emission limits scenarios (200 and 500 mg/Nm³ at 8% O_2).

The overall efficiency of Sorbacal®SP was compared to those of a standard hydrated lime and of sodium carbonate.

4.2.   Plant description

This float line is releasing about 70,000 Nm³/h of flue gases cooled from 450°C to 400°C before entering a contact reactor. The reagent is injected at the reactor bottom. Additionally, the facility is equipped with a 3-field electrostatic precipitator where by-products are separated from the treated gas.

The composition of the flue gases is as follows:
- HCl : ± 20 mg/Nm³ at 8% O_2
- HF : ± 15 mg/Nm³ at 8% O_2
- SO_2 : 800 mg/Nm³ at 8% O_2
- CO_2 : 10%
- O_2 : 8%
- Moisture : 10-15%

The plant storage and dosing equipment was dedicated to sodium carbonate, therefore a mobile dosing unit (Injecto-Matic® - See **Picture 2**) has been set-up on-site for the trial period (2 months).

Picture 2. Mobile dosing unit, Injecto-Matic®

4.3. Results

4.3.1. HF removal

The trials prove that Sorbacal®SP can effectively meet the HF emission limit even at very low feed rates. Specifically, at an injection rate of Sorbacal® SP at 10 kg/h leads to an HF concentration lower than 2.5 mg/Nm³ at 8% of O_2 whereas a feed rate of 40 kg/h for sodium carbonate was not enough to meet the HF emission limit of 5 mg/Nm³.

4.3.2. SO₂ removal

Results shown in **Figure 5** demonstrate that the desulfurization performance of Sorbacal®SP and sodium carbonate are, in these operating conditions, equivalent. Additionally, the SO_2 emission limits of 500 and 200 mg/Nm³ at 8% O_2 *i.e.* SO_2 removal rates of 38 and 75% are easily reached with 40 and 120 kg/h of Sorbacal®SP, which correspond to stoichiometric factors of 1.8 and 2.7. In the same experiment using standard hydrated lime, a feed rate of 40 kg/h was only able to achieve a removal rate of 30%.

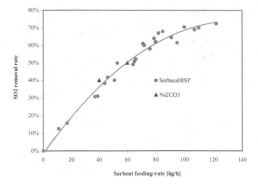

Figure 5.   Application trial in the glass industry.   Desulfurization performance of Sorbacal®SP and sodium carbonate as a function of the feedrate

5.   CONCLUSIONS

The parametric pilot scale study demonstrates that below 450°C, SO_2 removal performances of Sorbacal®SP are significantly better than those of standard hydrated lime.   However, beyond this temperature, the use of standard hydrated lime is recommended.   It also points-out that moisture improves the removal of SO_2 at temperatures below 200°C.   Additionally, utilizing Sorbacal®SPS instead of Sorbacal®SP leads to an increase of 15 to 20% of the desulfurization performance, especially between 150 and 350°C.

The application trial in the glass industry presented here confirms the increased performance of Sorbacal®SP over standard hydrated lime and more importantly, sodium carbonate.   Actually, the HF emission limit (5 mg/Nm³ at 8% O_2) is met at very low Sorbacal®SP feed rates.   Additionally, it has been proven that injecting Sorbacal®SP achieves the required SO_2 removal efficiencies.

Using Sorbacal®SP and Sorbacal®SPS for dry flue gas treatment in the glass industry enables this industry to meet required acid gas emission limits with minimal reagent consumption and reduced residue production.

6.   REFERENCES

This document is largely inspired from the original article : "Traitement des gaz de verreries au moyen d'absorbant calciques »Published in Verre Vol.13 n°1 , January 2007

[1]   Brown C., *Pick the best acid-gas emission controls for your seedling*, Chemical Engineering Progress, (1998), 63-70
[2]   Walled G., Lallai A., *Reaction kinetics off gas hydrogen chloride and limestone*, Chemical Engineering Science, 49 (1994), 4491-4500

[3] Pettiau X., Francoise O., Gambin A., Laudet A., *evolution of the calcic absorbents for the treatment of gases by dry roads: Spongiacal®*, Purification of the gas effluents, Mons, (1997), A7

[4] Walters J.K., Akosman C., *The removal off HCl from hot gases with solid absorbents*, High Temperature Gas Cleaning, 3rd (1996), 415-425

[5] Shiguang X., Dezhen C., Changshun C., Hesheng Z., *Municipal solid waste incineration treatment and dry removal off HCl from its exhaust gas*, Energy Approximately., Proc. Int. Conf. (1996), 719-726

[6] Gambin A., Francoise O., Laudet A., *Treatment of acid gases by dry roads. Performances of a calcium hydroxide with high porosity*, Genius of the processes, Nancy, 2001

[7] Mereu F., Moreschi R., Scalett B.M., *Lime in the knell emissions neutralization and recycle off filter dust into the melting process,* International Knell Newspaper, 101 (1999), 39-42

TO WET OR NOT TO WET – THAT IS THE QUESTION – PART A

Douglas H. Davis and Christopher J. Hoyle
Toledo Engineering Co., Inc.

I. ABSTRACT

The authors examine the arguments for and against the use of water-wetting of glass batch. The argument "against" is the added energy cost (and carbon cost) to evaporate that water from the batch. The arguments "for" include cost-avoidance from raw material segregation during transport, loss of material as dust during transport, shortened furnace life from batch dusting and carryover, and reduced efficiency and glass quality from diminished batch circulation and melting in the furnace. Unless steps have been taken to minimize these problems, going to dry batch would be, in total, a costly change. This would be especially true for a float glass operation, where very high quality is demanded. Use of wet sand use is encouraged where practical to save the cost of drying.

II. INTRODUCTION

There has been recent discussion about water-wetting of glass batch; most from operators reviewing recent heat balances on their furnaces and noting the separate category "Water Loss". The money spent to drive off the water in the batch is significant.

The water traditionally added to glass batch brings a number of benefits. These include minimizing the loss of mixed batch as dust during transport to the furnace, minimizing segregation during this transport, reducing carryover and dusting in furnaces, and facilitating batch recirculation in the melter. For large regenerative furnaces, where long furnace life is critical to profitable operation and log formation at the charger is well-developed, TECO makes the argument that water-wetting is money well spent. For other furnaces the case may be less clear.

The costs and benefits of water-wetting need to be reviewed; old practices need to be held up to the sunlight. Just because the old-timers did it does not mean the practice should continue. However, there probably were some reasons.

III. DOES IT COST MONEY TO EVAPORATE WATER? CERTAINLY!

There is no "free lunch". Water added to the batch will indeed require energy to dehydrate the soda ash, to volatilize the water and then heat that steam to the temperature of the combustion gases. A portion of that energy will be returned to the process in the preheating of the combustion air, but at the end of the day, batch wetting represents an increased energy bill.

In Europe and other parts of the world, it is reasonably common for glass plants to receive their sand damp, having only been gravity-drained after processing. For such plants, this discussion of dry versus wet batch is moot. In the US, sand is nearly always delivered to the glass plant in the dry state, having been dried following wet processing. This is partly due to longer shipping distances making it impractical to ship water, partly due to historically lower energy prices in the US, and partly because of the inconvenience of handling wet sand and the expense of upgrading the storage and handling systems. Also, cold winters rule out a number of sites. When dry sand users wet their mixed batch and then drive off the water in melting, it is the second load of water that had to be evaporated. The use of wet sand should be encouraged where practical.

Table I presents the range of energy required for water evaporation and heating this steam to the exhaust temperature for various furnaces. Included is a 700 tpd regenerative float tank, a 400 tpd regenerative end-port tank and a 300 tpd oxy-fuel furnace. Shown for the regenerative furnaces are both the new condition where exhaust temperatures are low, and the late campaign or "experienced" tanks where the regenerator efficiency has decreased. Also included are various levels of water-wetting, from 3 to 5%. For float furnaces, cullet use varies from 15 to 25% of the total glass production. However, for end-port and oxy-fuel furnaces, assumed to be for container or fiberglass production, the cullet amounts can range from 20 to 60% of the total glass production. The annual costs presented here are based on an assumed natural gas price of $7.50 /MBtu and a bulk oxygen price of $0.14 / 100 ft^3. Lower gas prices weaken the argument for dry batch.

Furnace	T/d	Cullet	Effic	Exhst	3% Water	3.5% Water	4% Water	4.5% Water	5% Water
Float – Old	700	15%	67%	650 C	$288,000	$336,000	$384,000	$432,000	$480,000
Float – Old	700	20%	67%	650 C	$271,000	$316,000	$362,000	$407,000	$452,000
Float – Old	700	25%	67%	650 C	$254,000	$270,000	$339,000	$381,000	$424,000
Float – New	700	15%	72%	450 C	$238,000	$278,000	$317,000	$357,000	$396,000
Float – New	700	20%	72%	450 C	$224,000	$261,000	$299,000	$336,000	$373,000
Float – New	700	25%	72%	450 C	$210,000	$245,000	$280,000	$315,000	$350,000
End Port - Old	400	20%	67%	650 C	$104,000	$121,000	$138,000	$156,000	$173,000
End Port - Old	400	40%	67%	650 C	$78,000	$91,000	$104,000	$117,000	$130,000
End Port - Old	400	60%	67%	650 C	$52,000	$61,000	$69,000	$78,000	$87,000
End Port - New	400	20%	72%	450 C	$92,000	$107,000	$123,000	$138,000	$154,000
End Port - New	400	40%	72%	450 C	$69,000	$81,000	$92,000	$104,000	$115,000
End Port - New	400	60%	72%	450 C	$46,000	$54,000	$61,000	$69,000	$77,000
Oxy-Fuel	300	20%	70%	1400 C	$153,000	$179,000	$204,000	$230,000	$255,000
Oxy-Fuel	300	40%	70%	1400 C	$115,000	$134,000	$153,000	$172,000	$191,000
Oxy-Fuel	300	60%	70%	1400 C	$77,000	$89,000	$102,000	$115,000	$127,000

Table I – Annual Out-of Pocket Energy Cost of Batch Wetting

Figure 1 presents annual out-of-pocket costs for older 700 tpd float furnaces (lower efficiency and higher exhaust temperatures). The costs (potential savings) vary from nearly $500,000 down to $250,000 annually.

Table II is a simplified look at the out-of-pocket costs for using water- wetting. This is a simpler view of the potential savings. The largest savings are for batch with maximum water and lowest cullet use. Batch with minimal water and high levels of cullet offer the least potential savings.

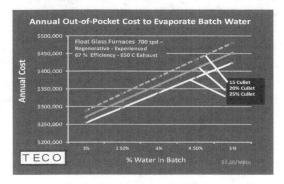

Figure 1 – Annual Water Cost for Experienced Float Tanks

Table II Annual Out-of-Pocket Costs for Batch Wetting	
700 t/d Float - Experienced	$254,000 - $480,000
700 t/d Float – New	$210,000 - $396,000
400 t/d End-Port - Experienced	$52,000 - $173,000
400 t/d End-Port - New	$46,000 - $154,000
300 t/d Oxy-Fuel	$77,000 - $255,000

IV. DO GLASS PLANTS SEE AN INCREASE IN THEIR FUEL BILLS?

It is difficult to supply hard data here. Some of the current experiments on trying dry batch may indeed provide some data. Most glassmakers, however, have been using wetted batch long enough that their current operators and engineers do not remember trying to use dry batch. Many of the conversions happened during the 1960's-'70's when the US soda ash suppliers were very proactive in the area of segregation prevention and carryover reduction. Most of these conversions were to a 3-4 % water-wetting, and any increase in gas usage was apparently not considered significant in light of the perceived benefits, especially not having to run the furnace so hard to obtain the needed quality.

The potential savings from removing the water from the batch has been shown above. Our next step is to review the benefits that come with water-wetting and estimate the costs of giving up these benefits in favor of dry batch.

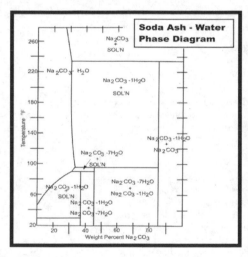

Figure 2 – Soda Ash-Water Phase Diagram

V. POTENTIAL BENEFITS FROM WATER-WETTING

A. Water-Wetting Provides Some Benefit in Meltability

1. Distribution of Fluxes

The initial water-wetting of the batch dissolves a significant amount of soda ash. Figure 2 shows the equilibrium phase diagram for a mixture of soda ash and water. The important word here is equilibrium, because the diagram shows the balance that must be in place in the end, but in the beginning it may be quite different. On the same phase diagram, shown on Figure 3, Line A would be the position of the soda ash / water mix on the diagram if all the water added to the batch got to all the soda ash. This would indicate there would be some permanent amount of residual liquid solution present in the mixed batch. However, since water added to the batch contacts materials other than just the soda ash, line B more likely represents the equilibrium ratio than does line A.

At equilibrium, there would be no liquid solution remaining; all the water present being held as solid hydrates of soda ash. This leads to the confusing terminology of wetted batch "drying out"- not really dry, but appearing dry.

In the batch mixer, of course, a lot of water is added to the batch all at once, and it takes a number of hours for this water to fully penetrate the soda ash particles and approach equilibrium. Therefore, for some period the temporary chemical balance is more like within Area C, with a large amount of liquid soda ash solution in the batch. This soda ash solution coats all the batch particles, including the sand grains, as shown on Figure 4. This brings the alkali fluxes into more intimate contact with the sand grains than would happen with dry batch. The cementing action of this coating also has some benefit in the melter, reducing dusting and carryover.

Chemistry does win in the end, however, with formation of solid hydrates and "drying" of the batch. For an old reminder, wetted mixed batch needs to be stored above 40°C, where the soda ash hydrates being formed are only single hydrates. This allows a number of hours of storage before the batch water transforms into crystalline hydrates and the batch gets really

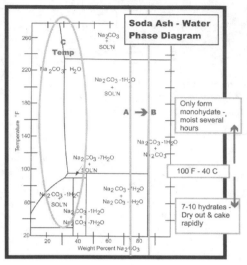

Figure 3 – Soda Ash Diagram

"dry". If the batch is allowed to become colder than 40°C, the need to form 7- and 10-hydrates kicks in and the batch "dries" quickly, with caking and plugging a problem.

A. Water-Wetting Provides Benefit in Meltability (Cont'd)
2. Fluxing Of the Glass from Added Hydroxyls?

One question to consider is whether batch wetting will improve glass melting via reduced viscosity. Hydroxyl ions that make their way into the glass do provide a fluxing effect. Such effects are maximized in oxy-fuel furnaces, where atmospheric steam content can exceed 60% and moisture in the glass can get to measurable levels, perhaps 0.125 or 0.250 wt%. But hydroxyls in the glass from water-wetting alone will be at a very low level. There should not be a measureable effect.

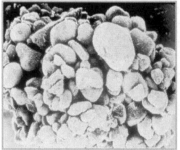

Figure 4 – Water Wetted Batch - Cold

3. Other Benefits in Meltability

Water has an impact on the dynamics of melting even at the cool temperatures within the batch piles. In the background information that accompanies their modeling software, Glass Services report that water has higher thermal conductivity than air which is displaced from batch by water vapor. Water vapor thus enables faster heating inside the batch. In addition, the evolution of steam has a mechanical mixing effect on batch, which causes a better contact of batch components and their mutual reaction. At higher temperatures, this mixing effect is carried on by carbonate decomposition. The practical experience is that moistened batch is melted faster than dry batch.

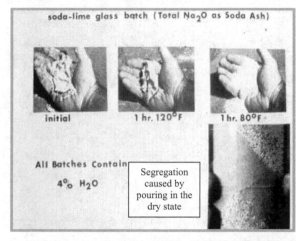

soda-lime glass batch (Total Na_2O as Soda Ash)

initial   1 hr, 120°F   1 hr, 80°F

All Batches Contain 4% H_2O

Segregation caused by pouring in the dry state

Figure 5 – Batch "Squeeze" and Dry Segregation

B. Segregation Is Minimized By Water-Wetting

In some cases, the benefits of batch wetting have been taken for granted. Water-wetting provides protection against potentially risky shortcuts in batch house design that have come into use for any number of reasons.

Literature dealing with batch segregation was more common in the 70's. But as recently as 2005, at this conference we have been reminded of the need to look for and alleviate segregation[1]. Water-wetted batch has come to be the traditional answer to segregation in handling. Preferably, water is added in the mixer, where the raw materials have come together for the first time. The water is added following complete blending, and the desired mix is locked in as long as the batch remains moist. The batch "squeeze" or "snowball" at the charger is normally our reassurance that the cohesion between particles has been adequate to prevent segregation of the mix and dust losses in the plant environment. This is shown in Figure 5. Even where batch moisture is minimized so that most of the added water becomes "lost" to hydration of the soda ash, the solid hydrate that is coating most of the batch particles helps to hold the mix together in spite of the work of gravity.

In plants not set up to handle and move wet batch, it is common to use wetting screws just prior to the charger, partially re-blending earlier segregation and giving a moist consistency for charging and for carryover reduction.

If batch wetting is not used and alternate approaches such as agglomeration are not used, there will be a cost. Dry batch will allow segregation in handling, in some plants worse than others. In TECO's opinion, there will be a decrease in glass quality or homogeneity. The options for dealing with this lower quality due to segregation are to 1) accept the lower quality, or 2) try to restore quality by going to higher furnace temperatures, or 3) by going to lower tonnages (longer retention) to allow residuals to melt and homogenize.

1. Payment For Segregation - Accepting Lower Quality
   It is unlikely that your customers will be willing to negotiate lower quality specifications, unless you are willing to lower prices. Putting lower quality out the door will almost certainly cost you customers.

2. Payment For Segregation - Higher Temperatures To Restore Quality
   The second option for dealing with lower quality due to segregation is increasing the melting temperature. If the possibility exists to safely increase the peak temperature of your melter, the cost in fuel to do so is estimated in Table III below. These costs are based on increasing the average glass temperature by 10°C, from 1490 to 1500°C. These added annual costs are significant, although smaller than the water evaporation savings previously discussed.

Table III – Fuel to Increase Glass Temperature 10 Degrees

Furnace Type	Additional Fuel M Btu / hr	Annual Cost 10 C Rise $ - @ $7.50/ M Btu	Water-Wetting Cost – Annual
700 t/d Float Regenerative	1.3	$85,000	$210,000 - $480,000
400 t/d End-Port Regenerative	0.58	$38,000	$46,000 - $173,000
300 t/d Oxy-Fuel No Regeneration	0.2	$18,000	$77,000 - $255,000

This higher glass temperature, however, will certainly bring another cost, i.e. increased melter corrosion by the hotter glass. The data on corrosion of AZS refractories, published by SEFPRO (Figure 6 below), shows that the common AZS refractories used in furnace sidewalls will have the corrosion rate increased 10-15% by the 10°C increase.

Although metal line cut can be alleviated by overcoating, there is a limit to how many overcoats can be applied. Of course while the overcoats help the metal line cut, the original blocks are getting thinner all the time. Fuel consumption is increasing with sidewall wear, but ignoring that factor, a 10% reduction in furnace life would be a reasonable expectation.

Table IV shows the added annual accrual that

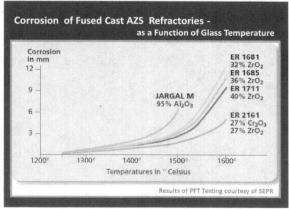

Figure 6 – Effect of Glass Temperature on Sidewall Corrosion

needs to be added to be ready for a rebuild at the shortened time table. The choice of a rebuild cost is difficult, since they may vary +/- 30% depending on the specific situation. However, this estimate uses middle-of-the-road rebuild costs, does not include the cost of money, and does not factor in cost inflation. The annual addition to the accrual cost should thus be conservative.

Table IV – Added Annual Accrual Cost for 10% Life Reduction				
Furnace Type	Anticipated Campaign	Planned Annual Accrual	10% Life Reduction Added Cost Distributed	Annual Distribution Over Remaining Life
Float	15 yrs	$1,400,000	$2,100,000	$156,000
End-Port	10 yrs	$1,500,000	$1,500,000	$156,000
Oxy Fuel	10 yrs	$800,000	$800,000	$89,000

Table V combines the cost of increasing the glass temperature by 10°C (from Table III) and the expected reduction in melter life due to greater corrosion (from Table IV) and compares the sum of these two to the potential savings from removing water from the batch. For the end-port furnace, the added expenses are greater than the potential savings. For the float and oxy-fuel, there is still some potential savings in dry batch use (although greatly reduced) if cullet is at the low end, and wetting had been at a high level.

Table V – Savings for Dry Batch Versus Fuel for 10 C Increase + 10% Reduced Life				
Furnace Type	10°C Temp Increase	10% Reduction Melter Life	Added Costs – Annual	Water Savings – Annual
Float	$85,000	$156,000	$241,000	$210,000 - $480,000
End-Port	$38,000	$167,000	$205,000	$46,000 - $173,000
Oxy-Fuel	$18,000	$89,000	$107,000	$77,000 - $255,000

3. Payment for Segregation - Lower Tonnage to Improve Quality
The third alternative to segregation problems is lowering melter tonnage. This represents one of the biggest financial drawbacks from dry batching, not being able to produce glass of the required quality at the same production level or output that one could with the wetted batch. Reduced tonnage will normally improve defect levels from seeds or stones. Unfortunately, with float glass and other premium glasses, lower tonnage probably will not restore the optical quality caused by batch segregation. With the increasing optical quality demands being placed on finished automotive glass, starting with a mixed batch that is less homogeneous may indeed impact your participation in that segment of the market.

Table VI below shows the expected "cost" of this alternative response. This is based on a reasonable sales value of the glass product, reduced by the value of the raw materials the plant does not need to buy. Obviously, the value chosen for the lost sales can be debated, but the number here for even a very small reduction in tonnage could overshadow any potential savings from eliminating a water addition. The larger tonnage reduction, 10%, is a common plant response required to satisfy a high quality market, and the potential lost profits here completely dwarf the savings from eliminating the batch wetting.

Table VI – Lost Income from Tonnage Reduction due to Segregation				
Furnace Type	Reduction in Tonnage	Sales Value / Ton	Lost Income / Year	Water Savings – Annual
700 t/d Float	10 tons	$350/tn - $75/tn	$1,004,000	$210,000 - $480,000
	10% - 70 ton	$350/tn - $75/tn	$7,026,000	$210,000 - $480,000
400 t/d End-Port	5 tons	$250/tn - $75/tn	$319,000	$46,000 - $173,000
	10% - 40 ton	$250/tn - $75/tn	$2,555,000	$46,000 - $173,000
300 t/d Oxy-Fuel	5 tons	$250/tn - $75/tn	$319,000	$77,000 - $255,000
	10% - 30 ton	$250/tn - $75/tn	$1,916,000	$77,000 - $255,000

C. Wet Batch Will Reduce Carryover Effects on Furnace Interiors and Regenerators.

Carryover and dusting is one area in which glassmakers are affected quite differently, depending on their raw materials. None are unaffected, except perhaps those using all-electric melters, but the impact varies.

Work by the GMIC to identify the survival needs of the glass industry highlighted the need to reduce the capital intensity of their operation. Obviously one approach would be smaller, higher-tonnage, less expensive furnaces. The Submerged Combustion Melter and the Stir Melter are possible candidates to fill the bill, but refining issues, among others, are not fully resolved.

Another valid approach to reducing the capital intensity is to extend the campaign life of the current furnaces. Most glass companies have been working hard to extend the life of their melters. Some float glass companies have exceeded 15-year campaigns without even checker repairs, with plans for much longer campaigns yet. The number of melt line overcoats is increasing and the measures taken to accomplish these melt line improvements are more extreme. It is TECO's contention that wet batching is a key tool in extending this life.

Carryover and dusting can be a serious problem inside the glass furnace. It can originate either from simple pickup of

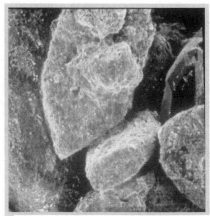

Figure 7 – Batch cemented by Hydrates

fine particles from the batch surface by the fires or from particles ejected into the sweeping gases by alkaline earth decrepitation. Contact of these materials with the interior refractory walls and crown can cause various direct deleterious reactions. Additionally, the interior surfaces can be simply eroded by rundown of glassy phases developed on the surface. This rundown not only damages the refractories but can give quality problems in the glass.

Carryover into the regenerators is a double-barreled problem. It is not only a problem for refractory wear, but seriously affects the regenerator effectiveness due to pluggage and surface loss. With some regenerators requiring an interim rebuild partway through a campaign, it is obvious that the furnace is already operating over most of its life with somewhat impaired efficiency. Regenerator problems are the predominant reason that a tank is shut down for a cold repair. Anything to minimize carryover is important for energy usage.

Water-wetting does reduce carryover and dusting. Once the batch is in the furnace, the cohesive moistness on the batch piles surface is gone. However, wet batch still acts to minimize carryover in the furnace due to the residual coating on the surface of the batch particles which connects many small particles as shown in Figure 7. This coating is from soda ash dissolved by the water addition in the mixer which is solidified by forming hydrates. Even after heating in the furnace, this coating helps to resist the prying effects of the combustion gases.

As we indicated at the start of this section, the degree of the carryover problem is highly dependent on the specific raw materials used at the plant. Two critical factors are 1) the degree of decrepitation of the dolomite and limestone used and 2) the fineness of the sand supply. Numerous factors determine the suitability of a raw material, including delivered price, but the two factors above are key to determining the amount of carryover that the use of dry batch creates. Simple XRF on samples from the top of the checker pack will tell you which problem currently predominates in your furnace.

The severity of the decrepitation effect, i.e. explosive decomposition of the carbonate mineral varies greatly between different deposits but also varies with the granulometry of the supply used. Coarser fractions show less decrepitation effect and this has led (particularly with dolomite) to using alkaline earth supplies that may be 16 mesh or coarser in spite of the segregation implications.

Choice of suitable glassmaking sand normally has to be a compromise. Some of the very fine sands in the central/western US areas are excellent choices for melting, but completely untenable with dry batch and some are not even suitable with wet batch. A TECO rule of thumb based on furnace life is that sands should have less than 7% finer than 140 mesh. Of course this is with the expectation of wetted batch. If a glassmaker chooses to use a coarser sand so as to make dry batch use more practical, the tonnage at which he can produce quality ware will be reduced. From the previous discussion, lost sales from tonnage reduction is a very expensive choice.

If a glassmakers' alkaline earth materials show low decrepitation and his sand of choice is coarse, there will be less need to protect against dusting and carryover. Dry batch will be less of a danger to the tank interior. However, many glassmakers are going to be hesitant to take chances with dry batch that is known to adversely affect furnace and checker life.

The financial cost of high carryover is similar to that discussed before on Table IV for reduced melter life due to increased glass temperature. This information is reformatted for convenience on Table VII below. Although the critical life factor would now be superstructure rather than tank walls, after a number of years this would still require a complete rebuild. A conservative estimate would be that a melter could have its campaign life reduced 10% by increased carryover and dusting. Using only the capital costs and ignoring the cost of money in a rebuild project makes the estimate even more conservative. The cost of the shortened melter life greatly reduces savings but may not eliminate them. Future carbon taxes will increase the savings from water elimination.

Table VII – Cost of Reduced Life Due to Carryover			
Furnace Type	10% Life Reduction Added Cost Distributed	Annual Distribution Over Remaining Life	Water-Wetting Cost – Annual
Float	$2,100,000	$156,000	$210,000 - $480,000
End-Port	$1,500,000	$167,000	$46,000 - $173,000
Oxy Fuel	$800,000	$89,000	$77,000 - $255,000

However, carryover and dusting can have a serious effect on the operation of your regenerators. Even if the economic choice of raw materials results in a better-than-normal situation for using dry batch, the regenerators are on a declining slope for efficiency. With finer sand and decrepitating dolomite, this decline in efficiency is going to happen much faster.

Predicting costs for this situation is difficult since damage to the regenerators can manifest itself in different ways. You may experience:

- Blockage of flues,
- Loss of air preheat temperature,
- Restricted flow calling for
  - Lower tonnage, or
  - Oxygen enrichment,
- Increased frequency of checker cleaning
  - Loss of pack due to thermal shock.

In any case, this faster degradation is at a significant cost to the overall operation. Table VIII shows the financial implications of reasonably small decreases in the combustion air preheat to a melting operation.

On Table IX below, the cost of a single 100°F downward offset of combustion air preheat is added to the 10% reduced melter life estimates from carryover and dusting in the melter. This is not particularly unusual even with water-wetted batch. A number of glass operations would find themselves handicapped with at least this much of a problem when switching to dry batch.

For the regenerative float and end-port operations, these two negative cost factors could readily overshadow any savings from removing the water and avoiding the evaporation load. Not all combinations of raw materials and firing would result in these performance-restricting factors, but without the protection of wetted batch, numerous operations would be negatively impacted.

Table VIII – Added Fuel Costs from Regenerator Damage (700 t/d float)			
Preheat Temp.	2200 F	2100 F	2000 F
Fuel use (scfh)	148,000	153,000	159,000
Gross M Btu/ton	5.6	5.8	6.0
Daily Cost @ $7.50	$27,400	$28,400	$29,300
Annual Fuel Cost	$10,001,000	$10,366,000	$10,694,500
Incremental/Yr	0	$365,000	$693,000

Table IX – Combined Costs of Carryover Vs Water Savings				
Furnace Type	Annual Addition To Rebuild Accrual Due to 10% Life Reduction	Added Annual Fuel Regenerator Damage 100 F Lower Preheat	Total Annual Cost of Carryover/Dust	Water-Wetting Cost – Annual
Float	$156,000	$365,000	$521,000	$210,000 - $480,000
End-Port	$111,000	$365,000	$297,000	$46,000 - $173,000
Oxy-Fuel	$89,000	0	$89,000	$77,000 - $255,000

Batch is Important to Effective Charging

The role of wet batch in charging depends on the charger style. With the screw chargers often seen on the smaller oxy-fuel furnaces or end-ports, the wet versus dry discussion should not greatly influence the pattern of batch logs on the glass surface; the logs being formed by breaking off from the end of the charged stream. The relatively small long-stroking chargers, sometimes oscillating, would be only moderately affected by dry versus wet batch, although the break between the individual logs will be less distinct with dry batch. However, the wide chargers used by the float glass industry depend very much on the steep angle of repose of the mixed batch at the hot face of the charger to provide their characteristic parallel log pattern.

As these pusher chargers retreat from their maximum extension, a "rotation" of the accumulated batch just inside the melter causes a significant split in the batch cover. With moist batch, this split remains distinct and defines the boundaries of the batch logs. If the batch is dry, the low angle of repose (flowable batch) will effectively fill the split with batch as the charger retreats. This has several effects. First, the big difference in height between the peaks and valleys is minimized or eliminated, giving less exposed surface area, and less room for melting "foam" to run and expose new batch to heating. This vitreous foam is not only an excellent insulator but an excellent reflector.

This flatter batch would be expected to require more area for melting or a reduction in the tons of batch to be melted. The added space may not be a major cost over the course of a campaign but is generally not available until a rebuild.

Secondly, the glassy space between "logs" is largely eliminated; giving less separation of the batch into discrete islands of floating batch that can move and respond to the underlying glass currents. The larger masses of batch will not be free to respond to the underlying rearward glass current and to move to the rear of the melter for a second chance at melting and homogenizing. More scum and silica relics will be carried on the end of the batch blanket up to and probably past the spring zone.

Losing the batch recirculation pattern results in the same need to respond to reduced glass quality that was discussed in connection with segregation, i.e. accepting lower quality, increasing the average glass temperature or reducing tonnage. As before, increasing the temperature is the most acceptable response, if the added temperature is available. If not, the use of dry batch is questionable.

VI. WATER-WETTING VERSUS DRY BATCH – A RISKY CHOICE

The potential risk in switching from water-wetted batch to dry batch depends on your particular situation, i.e. melter type, glass quality requirement, raw materials, charging technique, and as in most things, your fuel cost. The operations that spend the most on water, float glass operations, are most at risk for losing the benefits of water-wetting. These operations have the most costly melters, longer campaign lives, extremely high quality requirements, most batch exposure, and dependence on long parallel logs from pusher chargers. It is hard for us to think that a float glass operator would risk trying to retrieve the $250,000 - $500,000 a year by leaving out the water.

Table X summarizes what could be the potential costs of going to dry batch, comparing the sum of these against the energy savings in going to dry batch. In terms of segregation, the value used was the lowest cost of the three options discussed for trying to restore the glass quality from the effects of segregation. From the discussion of increased carryover due to dry batch, only the added energy due to regenerator damage was included. A shortened life on the superstructure only matches the shorter life on the tank walls from hotter glass responding to segregation. One rebuild will fix both problems. With respect to reduced glass quality due to charging problems on the float and end-port furnaces, increased glass temperature will be additive, as will the shortened melter life.

Where the risk incurred is less, with the smaller end-port, the potential savings is also less. But the prudent approach for both the float situation and the end-port would seem to be to retain batch wetting. With the smaller oxy-fuel melter, the choice is less obvious.

Table X – Summary - Potential Costs Versus Savings from Dry Batch					
Furnace Type	Segregation	Carryover	Charging	Added Costs – Annual	Water Savings – Annual
Float	$241,000	$365,000	$241,000	$847,000	$210,000 - $480,000
End-Port	$205,000	$186,000	$205,000	$596,000	$46,000 - $173,000
Oxy-Fuel	$107,000	0	0	$107,000	$77,000 - $255,000

Certainly one seed that has been firmly planted in the authors' minds during this rooting around, is, "Why don't more glassmakers use wet sand in the US?" Certainly we have many plants where freezing weather is a problem. The added cost of transporting water is a factor, as is the inconvenience and cost of handling wet batch. But if the energy spent on drying the sand at the vendor is eliminated and the savings passed on to the glassmaker, the energy savings justifies using the melter as a dryer and the glassmaker still has the benefit of a wet batch. This sounds like a win-win situation.

VII. THE NEXT STEP – PART B

This discussion has walked away (for lack of time and space) from several important possibilities such as:

- Alternatives to Preventing Segregation?
  - Viable alternatives to water for batch wetting?
  - Agglomeration, but with waste heat?
  - Changing your handling system?
- Alternatives for Reducing Carryover / Dusting?
  - Raw materials?
  - Agglomeration, but with waste heat?
  - Rapid fritting at furnace entrance?
- Alternatives for Creating a Recirculating Batch Pattern?
  - Alternate cullet and batch?
  - Modified charger designs?

These questions are under active discussion. They need to be answered, because batch preheating needs to happen, and then charging hot, dry batch needs to happen. TECO thinks answers are there to be used, but needing details to be worked out. But without these adjustments, switching to dry batch is a risky business. We will be back.

References

1). R.A.Barnum (Jenike and Johanson, Inc), "The Influence of Batch Segregation and Bulk Flow on Glass Quality", Proceedings of the 66th Conference on Glass Problems, October 24-26, 2005.

A HISTORICAL PERSPECTIVE ON SILICA AND THE GLASS INDUSTRY IN THE USA

Paul F. Guttmann
U.S. Silica Company
Berkeley Springs, WV

Recent reports in the media have pronounced the death of the glass industry in the United States. And if you are not close to the industry, you might believe these stories. The container industry in the USA in 1978 alone produced almost as much glass as all segments combined today. It is true that the glass industry has its challenges, but I believe that with challenge come opportunities. Glass manufacturers today have a global view of the market place and the glass industry remains alive in the USA as manufacturers participate in the global change that is occurring.

Production data for glass volume are difficult to find for the USA because data are not reported by any one government agency. Based on interviews with industry experts and secondary resources such as press releases and past DOE reports, I estimate the 2008 glass volume produced in the USA at 17.7 million metric tons, half of which is container glass.

The U.S. Geological Survey (USGS) estimates world production of industrial silica at over 120 million metric tons, 25% of which is produced in the USA. Total silica consumption in the USA, slowly declined from 28.9 million metric tons in 1978 to its low of 22.3 million metric tons in 1991. (Figure I) It then increased slowly over the next eighteen years to its current level of about 30 million metric tons. Glass manufacture in the USA consumes about 9.5 million metric tons of silica or about 30% of total annual industrial sand production. During the thirty-year period from 1978 – 2008, silica consumption by the glass industry declined 3 million metric tons or 25%. However, neither total silica consumption nor total glass production is indicative of the changes that have occurred in the various segments of the glass industry or in raw material supply.

Many observers might think little can be done with God's natural minerals like sand. The most pronounced change in raw materials for the glass industry came in the early '80s with the introduction of the quality culture. Quality programs emphasized by gurus like Juran, Deming and Crosby focused on processes rather than on the end product. This process focus led to improved mineral process equipment such as mineral separators and high intensity, rare earth magnets and thus improved material specifications. For glass manufacturers this meant finer top-size and fines control of silica for improved batch free time, new grades for specialty applications such as borosilicate glass and low-iron glass and improved control of ground silica for continuous filament fiberglass.

As I approached the retirement point of my career, I began contemplating the many changes I have seen over the my forty years supplying the 5 major segments of the glass industry and how raw material requirements have changed to keep pace with the industry. As an Illinois student in the 60's, I would stop into the Glass Problems Conference in the Illini Union and see the passion of the attendees for their industry. This passion has not diminished (even though the voices have softened) when I visited customers and the GPC today. The one constant in my forty years in the minerals business is change. When I graduated in 1968, I did not realize that I was witnessing the infancy of a new float glass industry or that I would see the end of the TV bulb business and the rise of flat panel TV's and something called "Gorilla" glass for my cell phone. For that matter, in 1968, I would have said "What is a cell phone?"

Before reviewing individual segments of the glass industry, we should briefly look at the economic times for the period covered by the data. During recession cycles, most industries are negatively impacted to some degree. (Figure II) We have had six recessions during the last 40 years and four of them were preceded by energy price shocks.

a. Dec 1969 – Nov 1970   A mild recession developed when the Fed attempted to close the budget deficits due to the Vietnam War.

b. Nov 1973 – Mar 1975. A quadrupling of oil prices by OPEC coupled with high government spending because of the Vietnam War led to stagflation in the United States. During this period we had an oil crisis with gasoline lines at stations, a stock market crash, high unemployment and frozen wages.

c. July 1981 – Nov 1982. In 1979, the Iranian Revolution created another Energy Crisis. The high energy prices created inflation, monetary tightening by the Federal Reserve and finally a housing recession.

d. July 1990 – Mar 1991. An expansion in the 80's led to inflation increases that again led to monetary tightening until a downturn in the economy. There was also another oil shock and monetary tightening from a Savings & Loan Crisis in the late 80's.

e. Mar 2001 – Nov 2001. After the longest period of growth in the United States, we had the dot.com bubble burst, a slow down in capital spending and the 9/11 terrorist attacks, which led to a recessionary period. Despite all, it was a brief period and many feel it might not have occurred without 9/11.

f. Dec 2007 – June 2009. This is the great recession due to sub-prime lending and the financial collapse of many institutions. We have all suffered during this one. The end has just been declared and although industries are starting to recover, many consumers and unemployed would not agree that the end occurred over a year ago. Economists who determine the official cycles said any further erosion would have to be in a newly declared recession.

CONTAINER GLASS

The largest glass-producing segment in the USA thirty years ago and still today is the container glass industry. This segment has also been the most dynamic in terms of volume shift during this time going from about 70% of all glass produced in 1978 to 50% of the estimated 17.7 million metric tons of glass produced in the USA today.

The history of the industry is captivating. Prior to 1900 glass containers were produced by artisan glass blowers. A proficient blower and four assistants were able to produce over 200 bottles in a 14-hour day. In 1903, Michael Owens produced the first automatic bottle machine and today some machines can produce over 700 containers per minute or over 900,000 bottles per day. By 1920, two hundred automatic machines accounted for 45% of total US production. In 1923, the Hartford-Fairmont Company (Emhart Corp) introduced worldwide the first gob shearing and feeding device with the first four IS machines installed at Carr Lowery in 1927. In 1929 the Owens Bottle Company, which was incorporated by Michael Owens, Edward Drummond Libbey and 3 associates in 1907, merged with the Illinois Glass Company of Alton, IL to become Owens-Illinois. The industry in the USA expanded to over 50 companies and then consolidated into the current 3 major corporations and several independents. In 1970, there were 120 glass container plants. Today there are 48 plants in 22 states. Competitive forces have driven innovation within the industry. But it has not been competition within the industry as much as competition from competing forms of packaging.

Steel cans were invented in the 18th century to solve the problem of keeping food fresh for Napoleon's soldiers in battle. It was not until 1935 that American Can solved the pressure issue of beer in cans by developing a coating to prevent the beer from reacting with the metal. But beer in cans became most popular in the late 60's when the two-piece, ring-top aluminum can became more competitive. Sealed waxed containers for milk were introduced as early as 1929, but while I remember them in the grocery store where I worked in high school, they never caught on with the milkman who made his home deliveries. Then in the 70's, the gallon plastic container for milk and the two-liter plastic container for soda took hold. From 1972 to 1983 the increase in glass containers was only 1%; cans were 1.7% and plastics were 9%. By 1990 plastics increased to 19% of all package types and by 1994 practically all soft drinks were in plastic. In 2001 Gerber Foods made the decision to package all its baby food in plastic containers, although today some is still packaged in glass.

What happened to the glass industry during this time? Was the media correct? Did the glass container industry die? No, the industry faced its challenges and found opportunities through marketing and technology. Although Owens-Illinois introduced a one-way bottle called the "stubby" as early as 1935, it was not until the 60's that the one-way glass container, with a hot end coating began a renaissance for the industry. This was followed by the two-liter, "plastishield", generic bottle in the mid-70's. Owens-Illinois initiated an extensive advertising campaign using the famed bowling announcer, Chris Schenkel, touting the "good taste of beer" in TV ads. The Glass Packaging Institute (GPI) followed with TV advertising and in 1983, partnered with labor unions to form the Nickel Solution Trust to promote glass containers and recycling. Glass marketing improved sales of wine coolers in the mid-80's, beer in longneck bottles, alcoholic spirit-coolers, flavored drinks and single serve juices and ice-teas in the 90's. In 1994, microbreweries became a major factor along with promotional beer bottles such as the Coors bat. By 1997, craft brewery sales reached 4.7 million barrels, almost entirely in glass. Designer appeal and the annual "Clear Choice Awards" by GPI have also kept glass in the forefront. Today new shapes and full body shrink sleeve labels enable containers to stand out on store shelves. In addition, new concerns on the health effects and recyclability of plastic containers, coupled with the organic movement have renewed the demand of containers. Once again we see milk in glass bottles with more products to come.

Technology also improved in the glass container industry. With the development of the one-way container and a hot-end coating to improve flaw resistance, manufacturers began the process of lightweighting containers by at least 30%. Manufacturers continue to work on improving the strength of glass in order to be more competitive. Increased use of cullet has also improved the competitiveness of glass. In addition to environmental benefits, cullet usage has reduced energy costs. In the early 70's, cullet use was insignificant. In 1985 it grew to 10% of the batch and doubled to 20% by 1990. Today the EPA puts the glass recycle rate at 28%. The container industry has set a goal of 50% cullet usage by 2013. In order to achieve this, the quality and quantity of cullet sources need to continue to improve.

Process technology has also improved the competitiveness of glass. Glass manufacturers and suppliers today focus on incoming raw materials and combustion controls. Capacity did not drop as fast as the plant closures would indicate because many of the plants closed were old and replaced with more productive, newer equipment in other existing facilities. An example of this is the 2005 Owens-Illinois Windsor, CO plant being the first container glass plant built in over 25 years.

What has change in the container industry meant for silica suppliers? Because of the lightweighting of containers and increased use of cullet, the decline in silica consumption has been greater than unit glass production. (Figure III) The graph would be even more pronounced if glass volumes were by weight rather than units. Silica suppliers have worked with container manufacturers to improve process economics by controlling consistency of raw materials. Process controls for silica include size equipment to limit the top-size of silica grains and the fine particles in the distribution. In addition, heavy refractory minerals are removed and often iron oxide is processed for consistency.

FLAT GLASS

The flat glass segment of the glass industry is an old industry that came alive with new history beginning in the 60's with the invention of the float process. In today's vernacular, the float process was a paradigm shift that in financial terms moved the needle. The industry has grown by using glass as a material rather than as a product. The industry has created many value-added products from this material called glass. I did not realize when I graduated that the float industry was less than 10 years old. Early history is interesting for flat glass, as many glass manufacturers have existed for over 100 years, an indication that they meet challenges by creating opportunities. Saint Gobain was formed in Europe during 1665 to produce mirrors. Pilkington, Corning, PPG Industries and Libbey-Owens-Ford, who produced the glass for the Empire State Building, were all formed in the 1800's.

In 1902, Emile Fourcault of Belgium developed the direct upward draw from the glass tank. In 1928 the Adamson Flat Glass Company in West Virginia, with an investment from LOF installed new Fourcault units valued at about $1million. That same year PPG mass produced sheet glass using their Pittsburgh process that involved drawing a continuous sheet of molten glass from a tank vertically up a four-story forming and cooling line. They also developed the Creighton Process to economically laminate glass for automobile windshields. From this, PPG introduced Dupate® laminated safety glass.

In 1952, Sir Alastair Pilkington invented the float process that enabled variable thicknesses. It was not introduced commercially until 1959. PPG Industries started the first float furnace in the USA in Cumberland, MD in 1963. The impact of the float process is seen in Figure IV. The float process took hold and by 1973, it accounted for 59% of domestic capacity and by 1984, essentially all 39 tanks in the USA were float. The industry has grown during this 40-year period to over five million tons of annual production. New technologies and processes have taken float glass across segments into the area of specialty glass such as solar applications.

Coincidental to Pilkington, Corning was developing a process to make flat glass and in 1966 patented their fusion draw process. This process was put on the shelf in favor of the float process but was later taken off the shelf and became the favored process for making LCD glass and is now Corning's vision for solar glass.

Value added products from float glass have been the key to growth. Double and triple pane insulating glass was introduced for energy savings. In 1983, Low-E coatings were introduced to provide cost effective triple pane performance in dual pane construction. This led to a growth period of increased use of glass in buildings. Other developments include self-cleaning windows in the early 2000's, improved replacement windows for remodel and solar control coatings. Increased vehicle usage came with new improved fabrication and curvature processes and the desire for larger vehicles such as SUVs and crossovers. Automotive coatings included "heads-up" displays, on-glass antennas and window defrosters.

The current growth for flat glass is in the area of high transmission, low-iron glass applications. Initial production of low-iron glass was utilized for niche markets such as furniture, high-end appliances and display cabinets. Today architectural designs and solar glass production are driving annual growth rate projections of 25% to 35%. Architects like the flexibility that low-iron glass provides in award-winning building design with its brilliance and clarity of the glass and truer color when coated. Architects find that low-iron glass can unite with a low-emissive coating to provide good environmental control of solar transmission. One example is the new Comcast Center in Philadelphia that is the tallest building in the USA to achieve the Leadership in Energy and Environmental Design (LEED) Gold Certification using low-iron glass. In the future, built-in-photovoltaic (BIPV) and organic light emitting diodes (OLED) are expected to increase demand in commercial buildings for low-iron flat glass.

While commercial construction is currently the largest end-use, low-iron glass used in solar applications is one of the more active areas of research and development. Solar applications include photovoltaic modules, solar thermal units and concentrated solar units. Several types of glass are used for these applications.

In photovoltaic modules, regular float glass is used as a back glass to which the cells are attached. A low-iron cover glass of high transmittance is used to protect the cells. The glass must be strong, chemically resistant to weather conditions and low in weight (i.e. thin). Over 90% of the solar rays can be transmitted through a low-iron glass and this can be improved by using thinner glass, patterned glass or anti-reflective coating that helps prevent reflectance of the rays.

Solar thermal units have similar requirements as photovoltaic low-iron cover glass. Concentrated solar power (CSP) units contain a system of low-iron glass tubes through which fluid passes and is heated by the solar rays. Concentrated solar power units require low-iron glass in both the mirrors and the tubes containing the liquid that is heated by the solar rays.

The solar market is one of high expectations and growth. Currently most flat glass companies, because demand does not equal a full furnace, campaign low-iron glass with regular glass. This transition time naturally increases manufacturing cost, but campaign times are becoming longer with increased demand. The high transmission glass market is one where silica companies can partner with glass producers. Although their natural deposits often limit mineral suppliers, if the deposit is conducive to processing the iron content of silica can be lowered. One specific case is the Rockwood Michigan deposit. For years this plant was considered for closure because its specialty glass base had declined. With the low-iron glass market surge, US Silica has just completed the newest low-iron silica plant in the USA by building a new process plant with a capacity of 500,000 tons per year to replace the former plant.

INSULATION FIBERGLASS

Corning Glass did the first commercial level research on fiberglass in the 1920's. This was followed by Owens-Illinois making glass fibers for insulation at their Newark, OH plant in the 30's. In 1935, Corning approached O-I with a proposal to collaborate on glass fiber development and in 1938 Owens-Corning Fiberglas was formed to further product development. (Perhaps the final driver for forming OCF was the formation of a subsidiary by Saint Gobain in Europe during the previous year dedicated to fiberglass insulation?) The first major application for fiberglass insulation was not in homes but in 1939 when the US Navy Bureau of Ships specified fiberglass in all new warship construction. Other applications such as bonded mat, battery separators, sewn blankets, air filters and staple electrical wire insulation followed during the war. Home insulation came with new construction following the war and Owens Corning was taken public in 1952.

Today an estimated 2 million tons of fiberglass insulation is produced annually in the USA. (Figure V) A major challenge for the industry came in 1994 when fiberglass became listed as a suspected carcinogen but the industry met the challenge by working with their trade association and developed scientific data to successfully have it lifted. New products have recently focused on the environment and energy savings. Industry silica data does not correlate well with insulation production because of several reasons. The industry data reported by the USGS is suspect because silica is often supplied by non-reporting local producers and thus estimated by the USGS. The government does not report production data for fiberglass production and thus it is only estimated in purchased research reports or by internal sources and not published. Finally, the major change in fiberglass has been the increased usage of cullet that eliminates the use of silica. Cullet as a major ingredient has been occurring since the late 80's. By the mid-90's usage was estimated to be 600K tons, mainly plate glass cullet. In 1992 the North American Insulation Manufacturers Association (NAIMA) began a formal recycling program. They announced that members producing insulation in 2006 used 2.1 million tons of recycled materials, which includes slag for rock wool. This compares to 1.3 million tons used in 2004.

The mineral challenges for insulation fiberglass are less than for other glass segments and are usually placed more on the batch chemists at the glass manufacturers, who are cognizant of the need to constantly search for lower costs. As mentioned earlier, local sands, such as feldspathic river sands are often used with the main criteria being consistency in chemistry and being free of heavy refractory minerals.

TEXTILE FIBERGLASS

This industry segment really began with the formation of Owens-Corning Fiberglas by Owens-Illinois and Corning Glass Works in 1938. That was the same year that patents were awarded to Owens-Illinois for continuous glass filaments, textile material and glass fabric. Although fiberglass today is used mainly in plastic composite products, the material is still often referred to as "textile fiberglass." By 1944, OCF developed the first glass fiber reinforced boat hull and two years later came the first surfboard. In 1949, both PPG Industries and LOF acquired licenses from OCF. Three years later, Pilkington and St. Gobain were licensed by OCF.

Production of textile fiberglass in the USA is rebounding during 2010 after falling from its peak 2005/2006 levels of 1.25 million tons. (Figure VI) In the 50's, composites became viable to replace steel parts in automobiles such as the Chevrolet Corvette and in swimming pools and underground tanks. The development of chopped fiber strands led to the development of fiberglass-asphalt shingles. Since 1978 the market has more than doubled with the development of new applications. The goals to improve fuel economy increased the usage in automobiles and the push for alternative energy sources now has the industry focused on wind turbines. The USA currently has about 35,000 MW of installed wind power with 28,000 MW installed in the last five years. Research is concentrated on making larger turbine blades and on offshore wind farms.

Growth rates for the industry are projected at 7 – 9% annually with higher rates projected in the newer applications of energy and water distribution systems. While production rates have increased in the USA, new furnaces are not expected. This is a case where the US companies have maintained a global perspective by adding capacity in the Far East. They have done this because of the strong penetration of foreign imports from independent Chinese producers since the 90's. In 2009, an extraordinarily poor year for production in the USA, imports accounted for 20% of total demand.

Ground silica is used for production of fiberglass reinforcements in the USA and Europe because the batch must be fully melted when passing through micron-sized openings in bushings. Silica producers have worked with glass producers on quality processes and process equipment that have improved the particle size distribution and lowered production costs for both the raw material and finished product.

SPECIALTY GLASS

Specialty glass, as the name implies, represents the smallest segment by weight, but the highest unit value of glass produced. Specialty glass includes many sub-segments characterized by application, properties, composition and processing. Most consumers think of market application. When mentioned, we all have an image of stemware, optical lenses, LCD televisions, art glass, lab ware, lighting, etc. To the applications engineer, specialty glass is characterized by the properties it provides such as clarity, durability and strength. And to the batch and furnace engineer, specialty glass is most characterized by the glass composition and processing required. Major compositions are soda lime, borosilicate, high alumina, and phosphate. These compositions have different melting paths and even within a class, melting paths vary. The Department of Commerce reports production data in terms of value rather than terms of volume. Estimates for pressed and blown segments in 2008 were $4.1 Billion as a reference.

Silica data used for specialty glass is highly variable from year to year, but looking at silica and soda ash consumption trends leads to several conclusions. (Figure VII) The industry was in a decline after the energy crisis of 1979 and continued to decline until 1987 because of pressure from foreign imports. The industry then began to improve and in 1991 there were in excess of 35 plants (10 majors) in the USA producing specialty glass of various applications. During this time the color televisions with larger screens were increasing with the last TV tube plant being built in 1997 by Sony and Corning American Video in western Pennsylvania. From 1998 to 2002, manufacturers of glass panels suffered from overcapacity and low prices and eventually the last 5 cathode tube plants gave way to LCD production, primarily in the Far East. Corning became the leader in flat panel display glass, partnering with companies in the Far East. Today they maintain one US plant in Harrodsburg, KY, which is now also the focus for their new venture in solar glass.

Because of the numerous compositions and processes for specialty glass, silica producers have had to concentrate on producing specialty grades, some times for limited production volumes. The main characteristics for silica in specialty glass are chemistry, particle shape and particle size distribution.

The first requirement of all glass sand is to have no heavy refractory minerals present. If refractory minerals are not removed from the sand they will remain and create defects or stones in the glass. When present in the ore, refractory minerals can be removed by traditional mineral processing techniques. The iron content of glass sand has become more important in recent years because of the growth in low-iron glass with high clarity. Glass sand iron content is influenced by the iron stains on the surface of the grains that can be removed through flotation or magnetic separation techniques. The limiting factor on lowering iron content is often the iron within the grains, which is dependent on the deposit itself.

Silica grain shape is a sand characteristic often overlooked in the melting process. In general, angular sand grains go into solution faster than round grains. Angular grains have more surface area for the flux solution to attack and dissolve more quickly. Grain shape also contributes to the ease or difficulty of cleaning or scrubbing grain surfaces. In the case of attrition scrubbing, angular grains are often more difficult for iron stain removal than round grain sand.

The top-size of particle distribution is important because it is a determinant of batch-free time. Most sand for soda-lime glass is desired to be minus 30-mesh with minimal amounts finer than 140-mesh. Borosilicate and high-aluminum type glass containing little flux must use finer sand. These glasses used for flat panels, dinnerware and solar glass desire grain sizes finer than 70-mesh or finer than 100-mesh with tight particle size distributions. The availability of the latter types has been a difficult match of technical attributes with low volumes and the desire for low costs.

CONCLUSION

The glass industry is not dead; it is alive! Growth has always happened in some segments and is happening today. The glass industry and mineral suppliers will continue to meet challenges that will lead to opportunities. Producers will continue to improve the usable strength of glass and will continue to improve emissions and combustion processes. I have had the personal satisfaction to interface with customers and our operations personnel to engineer unique sand gradations for new specialty borosilicate glass compositions. Mineral producers will continue to work with glass producers in the USA in search of opportunities to remain healthy.

As the traditional silica markets such as glass and foundry have declined in the USA and customers develop a global view, silica producers have had to focus on new applications within the shipping confines of their domestic silica plants for survival. One current example of silica producers finding opportunities is in the oil/gas industry. The increased emphasis of energy independence in the USA had led to technology breakthroughs in drilling for oil and gas. In particular, horizontal drilling and hydrofracing in the natural gas shale reserves in the South and in the Northeast has placed a tremendous new demand for round-grain sand. The sand is pumped into the formation to prop it open and let the gas or oil flow from the well. In the past, only large sized silica grains were used. Today various gradations that compete with glass sand are used and in large quantities for each drill hole. Figure VIII indicates the rapidity of this market growth. This has and could continue to lead to a shortage of round and sub-round grain glass sands because silica producers are able to achieve higher return from this application.

The one constant that has not changed is the passion for glass that I have found in the scientists and customers during my forty years. I am sure that I will miss this in my retirement. Now I know why I see so much gray hair or no hair at these meetings. It is this passion of the new generation that keeps my generation young.

FIGURE I
SILICA CONSUMPTION

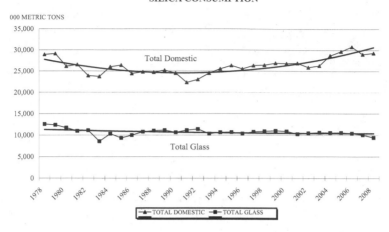

FIGURE II
CA Coincident Index (Fed Pa)

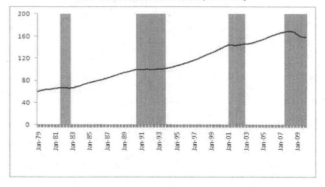

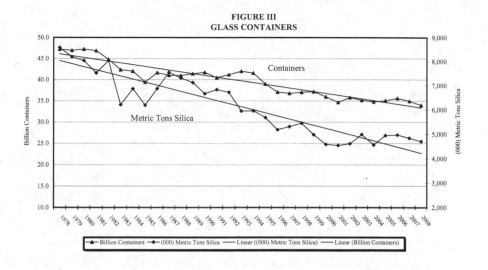

FIGURE III
GLASS CONTAINERS

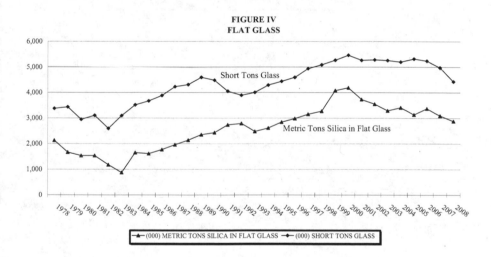

FIGURE IV
FLAT GLASS

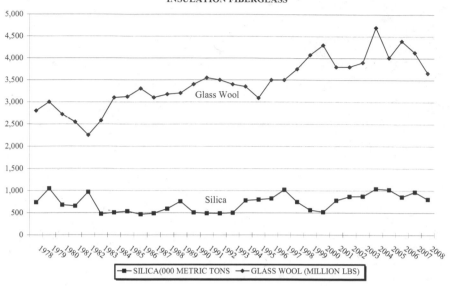

FIGURE V
INSULATION FIBERGLASS

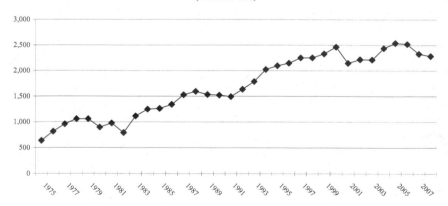

FIGURE VI
TEXTILE FIBERGLASS
(MILLION LBS)

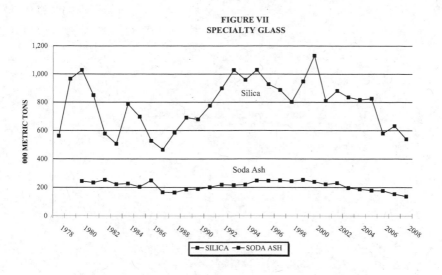

FIGURE VII
SPECIALTY GLASS

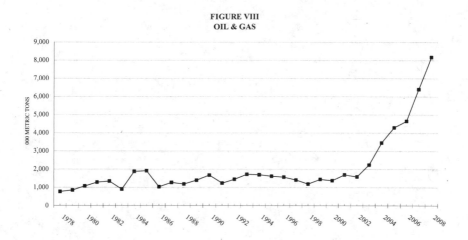

FIGURE VIII
OIL & GAS

Author Index